# EXPLORING ROLT'S LANDSCAPES

# EXPLORING ROLT'S LANDSCAPES

## WRITING, HERITAGE AND CONSERVATION

JOSEPH BOUGHEY

First published 2024

The History Press
97 St George's Place, Cheltenham,
Gloucestershire, GL50 3QB
www.thehistorypress.co.uk

© Joseph Boughey, 2024

The right of Joseph Boughey to be identified as the Author
of this work has been asserted in accordance with the
Copyright, Designs and Patents Act 1988.

All rights reserved. No part of this book may be reprinted
or reproduced or utilised in any form or by any electronic,
mechanical or other means, now known or hereafter invented,
including photocopying and recording, or in any information
storage or retrieval system, without the permission in writing
from the Publishers.

British Library Cataloguing in Publication Data.
A catalogue record for this book is available from the British Library.

ISBN 978 1 80399 271 6

Typesetting and origination by The History Press
Printed and bound in Great Britain by TJ Books Limited, Padstow, Cornwall.

Trees for Life

Dedicated with respect to the memory of
Brenda M. Boughey and Sara Richards for forty combined
years of love, companionship and support.

# CONTENTS

Preface and Acknowledgements — 9

1  Introduction — 13

    *Interlude  Around Banbury* — 29

2  Eighty Years of *Narrow Boat* — 34

    *Interlude  Cheshire* — 49

3  Influences and Involvements: Massingham, Organicism and Literature — 57

    *Interlude  Chester* — 73

4  The Early Inland Waterways Association and Waterways Revival — 80

5  To the Rural Idyll? Irish Waterways and Railways — 100

    *Interlude  Tardebigge* — 116

6  *Worcestershire*, Craft and Waterways — 124

    *Interlude  Llanthony* — 135

7  Encounters with Transport History: Willan, Hadfield and Rolt — 142

    *Interlude  Banbury* — 159

| 8 | Railway Revival, Conservation and Voluntarism | 166 |
| 9 | Afterword | 184 |

Glossary 188
Bibliography 189
Notes 192
Index 209

# PREFACE AND ACKNOWLEDGEMENTS

As I relate in Chapter 1, I have been interested in waterways and their history since infancy, and more generally interested in the natural and built environment and environmental politics and management since my youth. Between 1989 and redundancy in 2010, my teaching was partly around these latter areas, the most satisfying being the teaching of postgraduates – many of whom were professionally involved in environmental management and planning – at Liverpool John Moores University.

As I got to know the late Charles Hadfield in the late 1980s – writing an account of his work as *Charles Hadfield: Canal Man and More* – I became more aware of L.T.C. Rolt, who was a close friend of Charles after they met in 1946. I met his widow, Sonia – a very interesting person in her own right – after Charles' death in 1996. In the process, L.T.C. Rolt the writer became Tom Rolt the person – a warm, more human figure.

Re-reading much of Rolt's writings, it became apparent, as with Hadfield, that there was a great deal to discern, and much of his world that might continue to disappear. His thinking seemed to form an intangible heritage in need of conservation, even over aspects with which I (and others) would significantly disagree. I began to investigate Rolt's earlier life and to write and lecture about it, wondering about a biography. As with Hadfield, 'his life and work' seemed more accessible to write about than his personal life and, when he similarly had been involved in the making of history as well as dealing with the historical, it seemed best to focus on parts of that. This account thus concentrates on Rolt's earlier life and his involvement with waterways revival and the beginnings of narrow-gauge railway revival.

I am from a generation that could well have followed the suburban small-town life that my parents, post-war, might have envisaged for me. However, as with many from the middle class born in the 1950s and 1960s, I found myself seriously blown off course. In my case, it was an early commitment to environmental politics, partly spurred by an interest in waterways. Attendance at university and polytechnic also diverted me, as did two unconventional marriages.

What especially and personally blew me off course was the tragedy of losing my first wife, Brenda, in 2002 – which halted my research into Rolt – the subsequent confusion, and then building a new life with Sara, herself recently widowed and grief-stricken through this and other traumas. The long-lasting consequences of this were compounded when Sara herself passed away in April 2023. I had begun work on this book of essays – indeed, I recall reading through drafts while Sara slept during one of the many long visits to her in hospital – and she was pleased that I was working on it. I contemplated abandoning the project when faced once again with building a new life, but so much work had been done that I felt it best to continue. I must express my gratitude to The History Press, especially to Amy Rigg, in agreeing to a delay, without which I might well have been unable to continue. Thanks are also due to my agent, Georgia Glover, of David Higham Associates, for her caring support at a difficult time.

The essay form seemed to provide a useful way of tackling the many issues raised by studies of Rolt's work. Essays do not require comprehensive and exhaustive study, and enable close focus and depth with some issues, leaving others to be explored and followed up. I have emphasised particular issues and angles, some idiosyncratic, drawing on a range of sources. This has produced an advanced version of the inevitable conclusion that there is more to discover, explore, interpret, comprehend and conclude.

I have been an enthusiast for waterways history, in differing ways, for sixty years, and was an academic, teaching about the built environment, for twenty-one years. I have found the distinction between academic research and publication and (the best of) enthusiast research and publication tedious and counterproductive. Since retirement, I have been involved in adult education, with the Raymond Williams Foundation in the 2010s, the Merseyside-based Philosophy in Pubs since 2009 and, more recently, experiencing Gladstones Library at Hawarden. All three have emphasised the encouragement of scholarship, with learning as a mutual process, distant from the credentialled assessment-oriented learning that I experienced in university teaching. I have come across a sharp distinction between academic work in transport history and enthusiast work, and often with each side, despite many honourable exceptions, more interested in establishing and defending boundaries than in enlightening others. I hope that these essays can be read by academic and enthusiast alike: they are aimed at an intermediate level of knowledge and comprehension.

My thanks go to many people for a project that began over twenty-five years ago, and I can only mention some here. Chapters 2 and 7 are much-amended versions of articles that originally appeared in the *Journal of the Railway and Canal Historical Society*, for which the editor has kindly given consent. I would like to thank Stephen Rowson for the use of some of the historic photographs in his collection, and Pauline Soum-Paris of The Waterways Archive for her efforts in providing further historic images.

John 'Mellotron' Adams, and Claudia and Heather Wardle, kindly read and commented upon some of my essays. My sister, Dr Christine Barnes, was very helpful with genealogical queries. Tim Rolt, Tom's son, was exceptionally helpful in commenting on some obscure queries about his father. Victoria Owens kindly let me see part of the manuscript of her biography of Tom Rolt and offered helpful comments on my own draft text. In the now-distant past, Sonia Rolt was very helpful in recalling Tom and showing me round his family home in Stanley Pontlarge, along with showing me documents there that are now in the Ironbridge Reference Library and Archive.

I have benefitted from the help of archivists and staff at various places over the years, especially the anonymous ones at The National Archives, and a succession of archivists at The Waterways Archive, many of whom I have worked with as a volunteer. Voluntary colleagues at the archives and in the late-lamented Boat/Waterways Museum Society, who met regularly in the Rolt Centre – now the Tom Rolt Conference Centre – at Ellesmere Port, inspired me over the years with friendship and support.

Finally, but never last, my late wives Brenda and Sara always bore with my involvement in researching and writing about Tom Rolt and other subjects. I hope that they would be pleased that these essays have come to fruition.

None of these people and organisations necessarily endorse any of my views, and all errors are my own.

# 1

# INTRODUCTION

This book comprises a series of essays about the earlier work of the late Lionel Thomas Caswall Rolt, focusing upon his 'landscapes'. In this introduction, I explain my approach to some of the concepts and approaches involved, beginning with some of my own background.

While I was at primary school in 1963 and 1964, my parents introduced me to waterways holidays, hiring a boat whose journey included the waterways between Gailey, on the Staffordshire & Worcestershire Canal, and Barbridge, on the Shropshire Union Canal. This was less than twenty years after the publication of *Narrow Boat*, Rolt's first book, and we read about his trip as though it was a recent one. I recall a fascination with the Newport Branch, which led from the Shropshire Union at Norbury Junction, and his descriptions of this canal as barely navigable (by 1964 it had been closed for twenty years, although a society had been formed to restore it, which my late sister Hilarie, still at primary school, soon joined). Later, we acquired his *The Inland Waterways of England* and gradually realised that, by then, it was a portrait of the historical waterways of the late 1940s.

Rolt himself seemed to be a distant figure, who did not put too much of himself in his canal books; although later it would become clear how personal and particular his views were. By the time I learned of his early death in 1974, family canal holidays had largely ended.

It was later, when the controversies that led to Rolt's expulsion from the Inland Waterways Association (IWA) were highlighted in the late Ian Mackersey's *Tom Rolt and the Cressy Years* (1985) and then in the late David Bolton's uneven *Race Against Time* (1990), that my attention was drawn back to Rolt.[1] After I realised that much (then) recent history could soon be lost, I had begun to research the working life of Charles Hadfield, Britain's leading waterways historian. I became much more interested in Tom Rolt, who had been a close friend of Hadfield's, despite their significant differences in outlook. I met Rolt's widow, Sonia, at the time of Hadfield's funeral in 1996, and sought her help to recover more of Rolt's story and influence. I saw him as part of a group born in the Edwardian period, who had been influential in the making and exposition of transport history, and gradually investigated some of his disappearing world. After my first wife Brenda died in 2002, I set much of this work to one side, but interest slowly revived as the 100th anniversary of his birth approached.

I had worked on articles and lectures about Rolt and explored places that he knew, some of which were much changed. In 2011, I was approached to appear on a programme for the BBC: *The Golden Age of the Canals*. This involved a journey on the Shropshire Union Canal from Nantwich, north to Barbridge, following a route that my late father had taken in 1963, and, incidentally, one that Rolt had followed in 1939. Only when I boarded the hired boat at Nantwich did the filmmakers advise me that they wanted me to talk about Tom Rolt. Fortunately, I had engaged in a lot of research and reading about Rolt by then, although I had found myself developing ideas that contradicted some established accounts about him.

Later, I was more briefly involved in a sequel, *The Golden Age of Steam Railways*, in which, sitting on a pile of sleepers at Wharf Station in Tywyn, I commented on the Talyllyn Railway revival and Rolt's role. I was also involved in 60th-anniversary celebrations on that railway, taking part in a conference at Tywyn, at which I presented a paper to a largely railway audience, which formed the basis for a much modified article in the *Journal of Transport History*.[2]

In these investigations, Rolt seemed to be an ambivalent figure, exploring and admiring places and people at work, with ideas that diverged from much of mainstream politics and, perhaps, conservation. His writings inspired in me an appreciation of landscapes of various kinds, but also some doubt about their interpretation. Would I want myself or others to live in the kind of world that he desired?

Further considerations were raised by encounters with two places that were connected to Rolt.

## Mill Street, Nantwich

In the summer of 2021, seeking to minimise the threat of coronavirus during the pandemic that had begun early in 2020, I sat down with my late wife Sara to drink tea outside a bookshop in High Street, Nantwich. What had begun life as a bookshop with some catering was now a café with a bookshop: hospitality prioritised over the possibility of learning, perhaps. Looking down the north side of the High Street, I caught sight of a scene that seemed to evoke the world of Tom Rolt and display his influence. While the bookshop is flanked by the historic Castle Street, it was the three-storey buildings beyond – now shops with flats above – that I found more prominent. These were early nineteenth century, built from handmade bricks. The wall beyond the last one had featured two signs, a metal one and an older wooden one (which had partly crumbled away), both proclaiming 'Mill Street'. This had elements of something of which Rolt, who had come to know Nantwich in early wartime, would approve. He wrote in *Narrow Boat*: 'To step down from some busy thoroughfare onto the quiet tow-path of a canal [...] is to step backward a hundred years or more and to see things in a different, and perhaps more balanced perspective.'[3]

This street had led not to a canal but to a mill (Boughey Mill, as it happens) on the River Weaver, which is not navigable here; but in his time, this did indeed resemble a scene from a previous century. Nantwich had been a market town, specialising in Cheshire cheeses, with a widened High Street forming a market square in front of the present bookshop. Praising Cheshire farming and watermills, Rolt saw a place like Nantwich as the embodiment of organic relationships that stretched between town and country to relations between humans and machines. He also saw these as under threat, with little means of salvation. While he asserted that butchers sold only meat that had been locally raised and slaughtered, Cheshire cheese was being superseded by cheese that was an 'imported substance'.[4]

Walking round the corner into Mill Street, after a short pedestrianised section that is lined with older buildings, any sense of a surviving past diminishes rapidly. While some nineteenth-century buildings survive, the street becomes a modern road, serving a car park that is flanked by modern apartments, before a relief road, constructed in the 1990s, cuts across the final section of the street. The site of the mill, which burnt down in 1970, has been paved to provide a feature in a riverside park.

Back at the High Street, the Square has long been pedestrianised, with the older shops replaced by cafés, charity shops, chain bookshops, stationery shops and banks (the latter themselves now very much reduced). If Rolt evoked historic environments under threat of decay and destruction, what now remains provides only echoes of the past that he perceived. Ironically, the preservation of the bookshop building owes its survival to having been listed since 1948, under part of a planning system that he publicly abhorred. Several of the other buildings were listed in 1974, the year he died. He would, at best, feel ambivalence over such forms of protection, prompted by what the cultural geographer David Matless has called 'planner-preservationists'.[5]

To my mind, the opening of Mill Street provided a glimpse (and now only a glimpse) of what might be termed 'Rolt country': landscapes that he would recognise, containing elements of a past that had survived rather than being consciously recreated or conserved. These landscapes could exist in the mind, in clear memory or reconstructed through imagination, sometimes prompted by remnants and their interpretation. Rolt is perhaps less known or recalled for his interest in historic environments as for his emphasis on historic transport and, later, engineering biography, industrial archaeology and museums, the latter increasingly part of what would be later termed as 'the heritage industry'. He was prompted, however, less by antiquarian curiosity than by a desire to retain and revive elements of an apparently beneficial past social order whose declining traces he had perceived in the 1930s and 1940s.

He had a critical, practical involvement in two features of post-war conservation – the retention of Britain's navigable inland waterways and the preservation of narrow-gauge railways – but he sought to be a professional writer. The essays that make up this book focus on these roles, their background and the body of literature in which they were embedded. Rolt's landscapes could (and still can) be visited in the imagination as much as on the ground, but his interpretations can be clarified and contested.

A second beginning, seventy years after Rolt visited Nantwich, was an event I attended, which aimed to celebrate Rolt's life and work.

## Chester Centenary Rally, 2010

Historic transport featured at the Tom Rolt Centenary Rally of June 2010. Organised by the (then) Chester & District Branch of the IWA (an organisation that Rolt co-founded in 1946), this sought to celebrate the centenary of his birth. The event took place on the Shropshire Union Canal at the wide basin of Tower Wharf in Chester, which was able to accommodate numerous boats. Sixty years before, he had suggested the first such canal boat rally at Market Harborough in the English Midlands.

The rally's location was well chosen. Chester was where he was born, although his family had soon moved to Cusop, near Hay-on-Wye. Rolt first came to prominence for his writing about a voyage in the narrow boat *Cressy*, converted in the 1930s for pleasure use. *Cressy* was part of a fleet that was mostly built in the nearby Shropshire Union boatyard, which remained intact, if somewhat decrepit, by 2010 (it has since been somewhat revived). Like much that he favoured, it had survived through underuse and failure to invest in modernisation. A large plaque on the bridge opposite this yard proclaimed his connections with several causes that he inspired about historic environments.

The boatyard had been occupied by the firm of J.H. Taylor between the 1920s and the 1970s. This built flats and floats (large canal freight vessels) in the interwar period, and in the 1950s began to build specialised canal pleasure boats. Rolt had not intended it, but *Narrow Boat* had helped to inspire wealthier readers to consider acquiring, commissioning or hiring pleasure boats to open up and explore what he had depicted as a hidden world.

This did not include the basin on the towpath side of the canal, now flanked by tall apartments that mimic some of the lines of warehouses. There were canal warehouses here before the basin was filled in during the 1950s, but after 1921 they had not been served by the canal and, indeed, the re-excavated North Basin is still not used to moor pleasure boats. This reflected one of the fears of the early (and, indeed, later) IWA – that some water space might be retained for its appearance but not for any use for navigation.

The plaque is worth considering in detail. It disguises his later expulsion from the organisation that he co-founded, asserting: 'The success of the IWA in halting the dereliction of the canal system and ensuring its survival owes immeasurably to Rolt's vision.' The term 'immeasurably' provides a reminder that it remains unclear what would have happened after 1944 if *Narrow Boat* had not been published.

While dereliction was indeed halted, Rolt's vision only partly guided developments in waterways revival. These essays will attempt to reiterate, critique and conserve aspects of that vision. As he stressed historic conservation more than the development of recreational boating, much of Rolt's legacy for the current waterways scene is, perhaps inevitably, limited.

The plaque continues: 'His energy and the influence of his writings extending into the wide and varied sphere of what became industrial archaeology continue to enrich the life of many.' While the precise impact of any writing can never be finally known, Rolt's contribution beyond waterways probably enabled the enrichment of many more lives. My essays do celebrate his inspirational writings about (mostly) waterways and the enrichment of lives through the 'energy' of his practical involvement and commitment.

I feel that Rolt had two phases of interest: the first included his practical involvements, until 1953, with waterways and railways; the second, in which his somewhat apocalyptic pronouncements had been reduced, covered engineering biography and the growth of industrial archaeology and museums. These essays focus on the former period, but there are clear links between the two, including his membership and roles in the scholarly Newcomen Society, which he joined in 1945, and the Talyllyn Railway Preservation Society, which he co-founded and in which he remained active until his death.

A further plaque, by now one among several, was placed on his suburban Chester birthplace that summer. His widow, Sonia, although now frail, was able to attend the centenary event with the help of her son, and visited a stall for the IWA. On a platform opposite was a locomotive, *Tom Rolt*, which had been brought by road from the narrow-gauge railway in Wales that he had helped to revive, and his veteran car, an Alvis, which Tom's father had acquired in 1925.[6]

While at the rally, I began to reflect about what Rolt himself might have thought of this event. As a private man who eschewed personal publicity, he might well have been embarrassed by so much emphasis on his personal role. Would he approve of so many pleasure boats or preserved heritage vessels that had been restored but were rarely, if ever, used for their original purpose? Would he be keen to see the locomotive and car as exhibits, rather than in everyday use? And would he have been content with the loss of living landscapes that he admired – the Talyllyn and other preserved railways embodying heritage as tourism – or the lives upon carrying transport boats superseded by boats for leisure?

The desire to commemorate and commend much of his commitment and influence may have obscured the study of some of his ideas, especially the formative ones that took him into the 1950s. Reading these, partly through his autobiographies and in private letters, made me wonder whether the implications of these ideas were fully understood, and indeed about the lineages whereby sets of ideas and practices can be inherited in various manners. It also made me wonder whether, should his record of association with fringe right-wing causes become more prominent, this would, if left unconsidered, necessitate explanation and rehabilitation.

Commemorative events sometimes inspire renewed interest in a significant figure, with the formation of study groups and societies, conferences, seminars and journals. This has yet to take place for Tom Rolt, bar a website, which records tributes and lists his publications, some of which remain in print.[7] Rolt Memorial Lectures and the Rolt Fellowship, which began in the 1970s, continue, along with a major conference at Ironbridge in 2024. The 50th anniversary of his death has prompted Victoria Owens' biography, but otherwise there has not been much follow-up in research and writing; especially limited is any detailed critical assessment.[8]

## Appraisals and Influences

The 50th anniversary of Rolt's death in May 2024 and the 80th anniversary of the publication of *Narrow Boat* in late 1944 present opportunities for further appraisal of his work and influence. I would endorse the late Raymond Williams' view that a life can be continuous before and after the living of a physical life; in a secular sense, these essays concern influences before and beyond that life.[9] Rolt's influence does indeed continue, but in forms that maybe he could not imagine or endorse. We owe to him a broad shoulder upon the wheel of movements for waterways revival, largely for pleasure boating; for preserved tourist railways; for historic engineering, technology and industrial archaeology; and for a certain view of landscape as a living embodiment of an underlying sustainable society and economy. My main stress lies on the waterways and railway involvements, especially in writing, that brought about his practical engagements into the early 1950s. In many ways, he is a literary figure – no doubt a minor one, with a 'middlebrow' authorship and audience, but a literary figure, nevertheless. It is that figure, his actions and the views that he expressed and developed that are the subject of critique here. Tom Rolt the person, by most accounts friendly, charming and thoughtful, is not under appraisal.

Rolt's writings on landscape partly concerned processes that he felt had influenced or formed the places that he admired and whose conservation he sought in different ways. They also continue to draw attention to various such favoured landscapes. To revisit these is to attempt to perceive some aspects of what he saw and how much they have changed, but also sometimes to raise some doubts about his perspectives on these places. They may have also left unpreserved the forces that helped to produce and maintain these landscapes — the people, skills, crafts and meanings, artefacts, material culture, communities, homes and means of transport. Rolt, amid others, helped to conserve these through recording and describing — literary in his case, antiquarian or anthropological in others.

There are also the interior landscapes of his imagination. Visits to some places — many of them obscure — can bring out some elements of these, and I have incorporated a range of limited reflections upon recent visits, alongside more theoretical considerations. Composed in different styles, these form 'interludes' between the essays, but they might be viewed as brief essays in themselves.[10]

These essays comprise studies of Rolt's ideas and his work, but ones that depart from the somewhat semi-hagiographical assumptions that, perhaps quite properly, characterised the centenary events. I should stress my own perspectives, as some readers will disagree with the basis for my judgement. I am doubtful about much of Rolt's worldview, with its 'green' elements that could descend into reactionary romanticism.

Writing may assume an authorial consensus that readers share, but in this world of severe ideological divisions, even among members of the same middle class and supporters of neoliberal capitalism, this must now be untenable. For my part, I would turn round the oft-quoted assertion (attributed to the American theorist Fredric Jameson) that it is easier to imagine the end of the world than the end of capitalism. I feel that within the continued framework of neoliberal capitalism, it is difficult to envisage a position in which the end of the world for most of humanity has been averted. My comments, which differ greatly from Rolt's, should be viewed in this context; there are a range of alternative perspectives.

My own main influences for this book have been the reading of three academic writers: David Matless, cultural geographer of landscape; Patrick Wright, literature and cultural studies writer and broadcaster; and the late Raymond Williams, writer on cultural studies. None of these were contemporaries of Rolt. The oldest, Williams, shared a strong affinity with Rolt's love for the landscapes of the Black Mountains, which will be discussed later.

The others, in their differing ways, share an interest in landscapes, people and material culture from the past, often set in a context of literature. While Wright makes no explicit reference to Rolt, he discusses many figures with whom Rolt was loosely associated, mostly through H.J. Massingham and his associates, like Viscount Lymington.[11] These were, perhaps disconcertingly, on the fringes of the political Right between the 1930s and 1950s. Much of Wright's work is concerned with the analysis of places and their connections, often unexpected, with people; he is interested in the relations between the historic or 'heritage' and the national present.[12]

Matless, much more an academic specialising in cultural geography, discusses Rolt's *Narrow Boat* in his *Landscape and Englishness*. Perceptively, he places Rolt among agricultural organicists (several of them on the political far/fringe Right) and depicts him 'interweaving descriptions of humdrum objects with giant statements on civilisation'.[13]

What Matless calls 'canal culture', Rolt had sought through direct involvement rather than the external observation then emphasised and promoted by the Mass Observation movement. Rolt perceived an authentic folk culture in the painting of boats, a form of folk art, and collective singing in canal-side pubs like that at Shardlow in Derbyshire. He felt that these remains of a stable, organic, traditional England would not survive the attention of 'scientific planners', and Matless observes 'a fault line which over the following decades would come to dominate commentary on English landscape in an increasingly elegiac, melancholy, hopeless manner'.[14] That critical fault line, between the 'planner-preservationists' and 'organicists' will be explored in Chapter 3.

Matless stresses the contrast between Rolt's endorsement of the Potteries landscape as one truly reflective of the Industrial Revolution and the later modern factories of the Great West Road. Those who peopled a landscape are as significant as its appearance. Rolt felt that the boat people, or at least some of them, could be traced back to the 'pre-machine-age peasant', and suggested that they had links to Roma origins. Finally, Matless points out that *Narrow Boat* was unusual among organicist literature in having 'an immediate and considerable effect' in fostering the revival of waterways for leisure. This will be examined in later chapters, in particular Chapter 4.

All three writers display numerous differences with Rolt; it is doubtful that any of them ever met him, or that he would not disapprove strongly of authors who were more (Williams) or less (Matless and Wright) explicitly located on the political Left. However, all focus as he did on landscapes, 'heritage' and culture. While acknowledging the influence of broad economic social and political structures, all three consider linkages with and between more vernacular factors.

Rolt's writing and practical actions underlay two significant revivals – in British inland waterways from 1946 and in the conservation of narrow gauge and steam railways after 1950 – along with the appraisal and preservation of much of industrial archaeological and engineering history interest. These essays focus mostly upon the first two. Ironically, Rolt's very private character made him uncertain about the movements that he helped to start, and he would be in major doubt about, if not outright repudiation, of much that has been taken for granted in mainstream political circles since his time.

## On Landscapes

Despite the use of the term 'landscape' in two of Rolt's autobiographical volumes (the third was renamed *Landscape with Figures* posthumously), he did not define this, and it is a difficult and often contested concept. It may be appropriate to contrast Rolt's perceptions of landscape – partly as scenery, partly as an experience of nature sometimes approaching the mystical, and partly as a working landscape (he had helped out at a local farm near Cusop) – with those of Raymond Williams, who was born in the same Black Mountains area in 1921, the year when the Rolt family moved to Gloucestershire.

Like Rolt, Williams saw the writing of fiction as important, and *Border Country*, his first published novel, and his final posthumous novels, *People of the Black Mountains*, were set in the area. It is unlikely that Rolt would have read the former or would have gained much if he had done.

Williams grew up in Pandy, a small Welsh border village beyond the southern end of the Vale of Ewyas. His family had farming connections – his grandfather had been dismissed and turned out of his tied cottage by his farmer employer and became a roadman. It was a place that people left in large numbers, beginning with women going into domestic service in Birmingham or to the mining valleys of south Wales. Williams' mother was the daughter of a farm bailiff, while his father had started work as a farm worker before becoming a boy porter and then a signal worker on the Great Western Railway at Pandy.[15]

Williams described this as 'a very particular situation – a distinctly rural pattern of small farms, interlocked with another kind of social structure to which the railway workers belonged'.[16] He recorded that his father was friendly with local farmers and so they made up a close community, albeit one around acknowledged divisions. The Great Western Railway Company that Rolt would admire (and mourn), with its proud traditions, had been the same company that had victimised railway workers who supported the General Strike of 1926, including Williams' father. The railway workers, unionised and linked by the railway to industrialised south Wales, formed a separate body in this rural area, heading up the local Labour Party.

Williams was conscious of this as a community in which individuals supported one another, even when, as in his case, scholastic achievement meant that a member was to be sent away to university in Cambridge. He reacted to views expressed by Cambridge academics that Rolt might well have endorsed. He attended a lecture by L.C. Knights (co-editor of the Leavisite journal *Scrutiny*) on the meaning of 'neighbour' in Shakespeare, and when:

> Knights said that nobody now can understand Shakespeare's meaning of neighbour, for in a corrupt mechanical civilisation there are no neighbours, I got up and said I thought this was only differentially true; there were obviously successive kinds of community, and I knew perfectly well, from Wales, what neighbour meant.[17]

Williams perhaps idealised the community of his youth, but also acknowledged the continuing influence of its landscape. His character, Matthew Price, draws a significant contrast in *Border Country* when, studying historic population movements in the area, he finds the landscape and its distance in time and space intruding. His thoughts are relevant to Rolt:

> It was one thing to carry it in his mind, as he did, everywhere, not a day passing but he closed his eyes and saw it again, his only landscape. But it was different to stand and look at the reality. It was not less beautiful; every detail of the land came with its own excitement. But

it was not still, as the image had been. [...] He realized what had happened in going away. The valley as landscape had been taken, but its work forgotten. The visitor sees beauty; the inhabitant a place where he works and has friends. Far away, closing his eyes, he had been seeing this valley, but as a visitor sees it, as the guide-book sees it: this valley, in which he had lived more than half his life.[18]

Williams' view can be contrasted with that of Rolt: Williams saw the landscape as a product of labour and a place of work of various kinds. In thinking about landscape and going back, one can take the visitor's perspective, and in many ways this was true of Rolt. The labourers and craftsmen who built the much-admired Llanthony Priory in the Vale of Ewyas may have had aspirations but these may well have included getting paid and being able to find food and shelter. Aspirations may have rested with those who commissioned the building, although the short-lived monastic community there was seen as part of an initiative to counter Welsh insurgencies.

A later chapter returns to Williams and Rolt, but it may not be too reductionist to see their divergent views of landscape (and politics) as being conditioned by their initial class position and corresponding awareness of work and its shaping of environments. Rolt was drawn back in time, in place with Llanthony, by a partly mystical sense of belonging, and Williams too by a less-definable sense of rootedness. There is an element of self-rebuking in Matthew Price's feeling about looking at his home landscape as if it was from a guidebook, and perhaps this is an accusation that, if levelled at Rolt, would necessitate a robust response.

## Outline and Structure

The studies here focus on an earlier period of Rolt's writing and practice around seventy and eighty years ago, between the publication of *Narrow Boat* in 1944 and *Railway Adventure* in 1953. During the remaining twenty years of his life, he concentrated on writing professionally, some of it based on research. Committee involvements would partly reflect his earlier developed and stated views, but he diverged more and more into practical ends and achievement. 'Straight' biographical studies would *follow* him through these years but could not delve too far into his earlier world views, their contradictions, sources and implications. I have chosen to tackle this through themes rather than chronology, making this a series of linked essays.

This book opens with an essay that marks eighty years since *Narrow Boat*, Rolt's first and best-known book. It is set in the context of both earlier literary works and the waterways scene of 1939–40, contrasting this with the rapid critical examination carried out by Frank Pick of London Transport in 1941. The strong wartime influence of H.J. Massingham, a popular (and reactionary) rural writer, upon Rolt is explored over the development of his 'philosophical' work, *High Horse Riderless*, of 1945. This involved a rejection of industrialisation and called for a more rural, organic society, along with a stress on certain voluntary principles. Massingham is set in context with other organicist and rural revival writers and political figures, including Viscount Lymington, who would become a vice president of the IWA. The kind of nostalgic country writing of which Massingham was a leading proponent in the 1930s provides a background to Rolt's work and the perceptions of much of his likely audience.

Rolt did not write *Narrow Boat* in order to launch any practical campaign to revive inland waterways, but the book's impact after publication led him to co-found the IWA in 1946. While some of the IWA's early history has been detailed before, this account focuses on Rolt's role, some of the problems he faced and the significance of his involvement. In this respect, it diverges from some earlier accounts.

While he played no practical personal part in their revival, his portrait of some Irish waterways was recorded in *Green & Silver*. Based on a long voyage in 1946, this influenced the founding of the Inland Waterways Association of Ireland (IWAI). Rolt perceived special qualities in the west of Ireland, seeing it as a rural idyll, and commented on both waterways and railways. This commentary is evaluated against some of the realities of post-Emergency Ireland.

In wartime, Rolt lived on his boat *Cressy* at Tardebigge in Worcestershire, explored the county and wrote a volume on *Worcestershire* for a county series. He would distil his knowledge of inland waterways into a portrait of *The Inland Waterways of England*. Both books stressed the nature and significance of craft skills and people, and the desirability of their flourishing. The emphases in these portraits are considered, along with Rolt's early encounters with conservation.

Rolt's involvement with the historical has led some to regard him as an historian, but contrasts with two contemporaries – T.S. Willan and Charles Hadfield – indicate limits to his earlier historical investigations. His contribution to waterways history – with many insights, although perhaps with some missed opportunities – is discussed, making some differences in coverage and conclusions for the Worcestershire/Warwickshire Avon and waterways serving Leominster.

Rolt admired the railways before grouping and nationalisation, of which he disapproved, but developed the opportunity to revive one line that was not nationalised: the Talyllyn Railway in rural west Wales. This involved the pioneering of principles – voluntary ownership and management – which would eventually be applied elsewhere. The politics of conservation that this reveals are discussed. Unlike the inland waterways, Rolt would remain involved with this railway, despite its increasing dependence on tourism. His two years as railway manager, dealing with a distant board and local and national volunteers, are considered.

An Afterword attempts to draw together the discussions of Rolt's writings and his contribution to post-war conservation, both in practice and principle.

The longer essays are interspersed with short, illustrated essays – 'interludes', in Matless' apt concept – that discuss some of Rolt's favoured landscapes. These vary in style and approach, and each makes different points about places associated with Rolt that, however much changed, remain worth visiting.

The Tom Rolt Centenary Rally in Chester in June 2010 at Tower Wharf on the Shropshire Union Canal. The boat at the front centre is the largely rebuilt former inspection vessel *Lady Hatherton*.

The steam locomotive *Tom Rolt* was brought from the Talyllyn Railway to the Tom Rolt Centenary Rally and displayed on a low-loader.

Mill Street in Nantwich, July 2023, looking from the High Street at the junction with Mill Street.

## INTERLUDE
# AROUND BANBURY

Chapter II of *Narrow Boat* relates a road journey from Winchcombe to Banbury Cross in 1939, before *Cressy* was reconverted to a residential boat at Tooley's Yard. Characteristically, Rolt avoided the A roads, shown on his Ordnance Survey map as 'thick red lines as ugly as the roads themselves', and entrusted his route to smaller roads like the B4035. His stress on localism led him to commend the brickwork of houses in Upper and Lower Brailes, while those in the villages of Swalcliffe and Tadmarton were built of local limestone. Travelling this road today, these contrasts remain but the thatched cottages have been joined by much post-war housing. Self-sufficiency in local building materials has not been a feature for many decades.

Villages that served agriculture, much expanded, now serve commuters, retirees and perhaps increasing numbers of people who may work from home. These have become like the 'show village' condemned by Rolt, which, 'on deeper investigation, often turns out to be as lifeless as a stuffed bird in a museum, the cottages week-end dormitories for jaded business men, and the great barns riding-schools or Road Houses'.[19] Rolt was recognising processes that have accelerated to dominance in eighty years. One exception today might be social housing, at Tadmarton for instance, although his adverse remarks elsewhere show that he would not approve of this.

On 23 February 2018, my late wife Sara and I followed part of this route, imagining that Rolt would be satisfied with the continuing steepish gradients and bends in this road as it twisted round on course for Shipston-on-Stour. We stopped for tea at the Stags Head at Swalcliffe, a thatched pub (Grade II listed) with limestone walls. We asked the young barman what housing in the area cost – at least £350,000, was the reply, and that was before the house price inflation of the 2020s.

It was 2 miles further down, towards Banbury, that we saw what at first seemed like a strange Roltean moment. At a road junction with the B4035, what looked like a horse-drawn Roma caravan (vardo) came down the hill and swept into the lane leading to Swalcliffe Lea. Astonished, we followed this along, but when we drew close, we found that this was not a Roma van. A sign on the back proclaimed 'Sydney's Exploditions', and it was seemingly available to hire for a day out. The operator wore a hi-vis jacket with revolving warning lights on the vardo, but for a moment it seemed that time had not passed. It was a reminder, however, that what might have been a familiar enough sight in the late 1930s was now revived as a leisure device, like the 'shepherd's huts' that have never seen a shepherd but have become popular holiday lets.

The B4035 road ends in the suburbs of Banbury, 'besieged on all sides by semi-detached monstrosities'; Rolt later viewed a 'red rash of modern suburbia' from the church tower. Quite where he, with his deep suspicion of post-war planning, would hope that people would live is unclear.[20] Rolt could present a positive image of village communities, but details about one local village suggest a different picture – one of poverty that has now given way to affluence, just as with many villages within reach of generally well-paid employment.

Banbury features one local-based rural/urban enterprise, Hook Norton Brewery, which has been based in the nearby village of Hook Norton since the nineteenth century. A 1943 report into the village of Hook Norton indicates a great deal of change for that area, from a position that does not read like an unenviable decline since Rolt's time.

A contribution to *The English Rural Landscape* covered the history of Hook Norton, from which parts may be summarised. This was an 'open village', with no major landowner, something that Rolt does not seem to have considered elsewhere. Industries had come from outside – the brewery founded in 1849 but rebuilt by 1899 after the railway had come in 1887 – while for a period, from 1889 until closure in the 1940s, the Brymbo Ironstone works, processing iron ore, was served by rail. Many village buildings were built of local ironstone.

Kate Tiller's summary of the 1943 survey for *Country Planning* paints a gloomy picture:

> Here was a landscape of declining population, dilapidated agricultural buildings, and farming methods, fossilized by economic depression in a traditional mixed husbandry [...] Rural crafts had declined to the point that only three smiths worked in the area [...] Community activities were marked by apathy and lack of leadership. The centre of village life was the pub: Hook Norton had seven and a beer shop. The authors

argued that mains water, gas, electricity, metalled roads, a bus past the door and active citizenship should be brought to these neglected areas.[21]

Rolt would have commented that village revival was indeed needed, but on a different basis – reviving agriculture in a more sustainable way and not the affluence brought by incomers to a village in which less than 2 per cent work in agriculture. Nevertheless, mains water came only in 1955 and sewerage in 1965.

As Tiller would conclude in 2000, 'There are few dilapidated buildings to be seen in a landscape which is still recognisably the product of its long history'. Many older buildings have been preserved, now providing a prosperous village with few clues about its poverty-stricken past, which may have been more realistic than Rolt's somewhat romanticised view.

St Mary's Church in Banbury. From the tower, Rolt viewed that 'red rash of modern suburbia'.

The thatched roof and stone Stags Head pub in Swalcliffe, with houses in local stone beyond.

The view west of the church in Hook Norton in 2023.

In an Oxfordshire lane near Swalcliffe a horse-drawn vardo is now used for leisure.

# 2

# EIGHTY YEARS OF *NARROW BOAT*

L.T.C. Rolt's reputation rests partly on his role in post-war waterways revival, the preservation of narrow-gauge railways and forms of historic conservation. The former began eighty years ago with the publication of his first book, *Narrow Boat*. This book proved to be extremely influential: the waterways historian Charles Hadfield, who was never one to exaggerate, attributed to it the post-war revival of Britain's inland waterways that fostered his own involvement: 'little of it, except maybe one or two canal histories, would have happened had *Narrow Boat* not been written. The time, the need and the man came together to produce the book.'[1]

The 'time' and 'need', which indicate and explain why the book was influential, will be discussed later. My initial focus here is on Rolt's reasons for writing a book that would be commended in 1979 by a Newcomen Society contributor as 'of no great importance in itself but [...] the first work of a writer who I believe to have been one of the two major influences in changing public indifference to industrial history'.[2] It outlines some of the influences that brought this book into being, not least some of his experiences of transport, engineering and landscape up to the 1930s.

Later chapters will consider its surprising consequences for the post-war revival of waterways. However, *Narrow Boat* seems an unlikely candidate for a book that would unintentionally inspire a movement to revive historic canals and river navigations.

Such a book might be expected to extol the virtues of boating for pleasure, providing some practical guidance upon the selection and navigation of boats, and the routes and facilities available. It would portray the virtues of scenery and of waterside 'heritage', dressed up perhaps as 'olde England', set within a romanticised historical outline. The book would discuss the possible revival of freight transport and the potential to transform transport routes into linear leisure assets. It would consider and develop explanations as to how this potential for transport and holiday boating had not been realised in the early twentieth century. Practical measures would be advocated to secure inland waterways' retention for leisure and amenity, along with the encouragement of trade and perhaps enlargements to develop new traffics.

*Narrow Boat*, which comprises an account of a long journey in 1939–40 on Rolt's narrow boat *Cressy*, provides none of these features beyond minor details of the boat's conversion in an appendix. The book did not set out to do so; others would grasp possibilities that Tom Rolt did not anticipate, although he would add comments for the second edition of 1947. To explain the form of the book, it may be helpful to consider the author's development and perspectives, which will be highlighted further in Chapter 3.

## Encounters with Transport and Landscape

Rolt's childhood in border locations placed him very much on the fringe of middle-class society and very much on the edge of the national railway and road systems. The only urban area that the young Tom Rolt knew well was the small cathedral city of Chester, where he was born and which he continued to visit into later childhood. In early adulthood, he lived for some time in the Potteries but, despite being both repelled and fascinated by their landscapes, did not explore their history in detail.

Rolt would not feel that steam railways (especially over narrow-gauge lines), or indeed early motor cars, presented any intrusive threats to rural landscapes, and he would view canal engineering as harmonious with nature.[3] Yet he did perceive a transition from what appeared to be timeless rural landscapes through the changes of the Industrial Revolution into a modern era with potentially adverse impacts of industry, transport and engineering upon landscape and society. That historians could unpick and challenge many assumptions behind this perceived transition (many rural landscapes, for instance, were not timeless but the product of Victorian agricultural depression, rendering these an uncertain heritage) does not invalidate its force as one inspiration for *Narrow Boat*.

Rolt's career as a premium engineering apprentice brought him into contact with two more influences that would be reflected in *Narrow Boat*. One was the interest in practical craft, which began in 1926 with his first apprenticeship with agricultural engineers near Evesham and continued into the 1930s; he learned both practical skills and a respect for craft, especially craftsmen. Although he called himself an engineer, he was not a civil engineer in today's sense. John Bate, a long-standing Talyllyn Railway volunteer, who was its chief engineer until 1994, much later stressed that 'while Tom Rolt was the only member of the early Committee with practical locomotive experience, he had served his apprenticeship as a working fitter. It is evident from his autobiographies that he never undertook any theoretical study or engineering design work.'[4] This perhaps led him to admire craftsmen who could follow established procedures skilfully and sometimes improvise solutions, but would not be expected to take an overview and design whole processes.

His practical abilities enabled him to both work on the conversion of the boat for *Narrow Boat*, from a holiday boat to a home, and to appreciate the practical work that underlay historical construction and engineering. Like his uncle, Kyrle Willans, who secured his apprenticeships, he was attracted by craft skills, rather than the scientific abstractions and bureaucratic elements of professional engineering knowledge and practice – and not by any accompanying financial rewards or social position.

The second influence was derived from his first encounter with the Potteries, which he explored between 1928 and 1930 while he worked for Kerr Stuart, inter alia, builders of narrow-gauge locomotives. The contrast between industrial landscapes and 'the green England that I had known as a child'[5] fostered an interest in the Industrial Revolution and how it had changed the nature of work, places and landscapes. He would be less interested in detailed factual history than the role of practical people like engineers, alongside the apparent spiritual losses and degradation of craft that industrialism had brought and continued to bring.

There was little appreciation of the social and economic relations involved, bringing pressure to abandon industrial crafts. Although he soon appreciated the influence of transport development on the Potteries, canals received no special attention, even though both his lodgings and workplace were located by canals. His life there accounts for the extensive coverage of Potteries industries in *Narrow Boat*, while his sympathetic account of Potteries workers contrasts with the more cursory dismissals of other industrial areas and their populations.

## Waterways, Willans and Motoring

Tom Rolt's first real encounter with canals came with the same narrow boat, *Cressy*, which he would later own. Kyrle Willans had purchased this former Shropshire Union carrying vessel, one of the last this company had built, in 1917. Willans had it motorised and converted to a holiday boat for eight at Frankton, on the Welsh line of the Shropshire Union Canal.[6] He had planned to establish a small pleasure-hire fleet, a venture then unknown outside the various hire firms on the Broads and the Thames.

After Willans moved to the Potteries, Rolt helped him to move *Cressy* from Ellesmere to Barlaston in March 1930, and during that journey came to feel that the boat:

> ... did not intrude upon the landscape; she became a part of it like the canal itself. As I realised this, my consuming interest in engineering and my feeling for the natural world, which, since I had come to Stoke-on-Trent had begun disturbingly to conflict with each other, were suddenly reconciled and before we had covered many miles I had fallen head-over-heels in love with canals.[7]

While Rolt took part in a further trip to Blisworth in Northamptonshire in August, soon afterwards Willans sold the boat.

This reconciliation of conflicting concerns was reinforced by some of his reading, such as Smiles' *Lives of the Engineers*, from which *Narrow Boat* would quote extensively. He was also influenced by the popular rural writer, Harold John Massingham, whom he would get to know well in the 1940s; this will be discussed in Chapter 3. He began to see the 1920s as a passing golden age of motoring, involving very limited traffic over well-surfaced roads, threatened by the development of mass-produced vehicles, freight and passenger-traffic growth and congestion.

He helped to found the Vintage Sports-Car Club (VSCC) in 1934, to support and celebrate earlier vehicles and their owners. After he became part-owner of the Phoenix Green Garage on the busy A30 in Hampshire in 1935, his experience of the surrounding area led him to see 1930s roads as a dystopia that increasingly outweighed his admiration for craft-built vehicles owned by practical enthusiasts. That the growth in road-borne freight would severely affect the canals (and railways) he admired was not yet too clear, and neither was the influence of motoring and public passenger transport upon the countryside and leisure.

Rolt wrote later: 'Although my interest in railway and steam locomotives dwindled almost to vanishing point during the years I spent at the Phoenix Green Garage my interest in canals did not.'[8] At his urging, Willans had reacquired *Cressy* in 1936, and subsequent voyages on the Grand Union Canal had not disappointed.

Around 1936, Rolt began to write short stories for publication and conceived the idea that he could live economically on board, travelling and making a living as a writer.[9] After *Cressy* was moved to Tooley's Yard boatbuilders at Banbury in 1939, he acquired and converted it from a holiday craft to a travelling home for him and his wife-to-be, Angela.

By now, he had come to see the narrow lowland canals as 'so many secret ways leading into the heart of England, peopled by men who were themselves a part of the English tradition'. Rolt's journey 'seemed to me to fulfil in the fullest sense the meaning of travel as opposed to a mere blind hurrying from place to place'.[10] While in the Potteries, he had conceived that:

> There still existed substantial pockets of rural England where beauty survived substantially unsullied and unpolluted by urban man and where the way of life that had contributed to its beauty, though visibly failing, still possessed sufficient tenacity and strength to give an eloquent meaning to the landscape. In other words, man was still playing his true creative role in the ecology that had produced the beauty.[11]

'Ecology' was then a term in little use and his stress was on the conservation of rural places, crafts and ways of life, specifically landscapes that he felt reflected a balance between nature and human impact and involvement. This idea will be explored in later chapters.

While he had then felt that the urban industrial Britain and the older rural world would coexist, he began to view the latter as under threat. His journey for *Narrow Boat* sought to travel the Midlands canals in search of this world, aiming to explore a past rooted not in nostalgia but in authenticity.

In this sense, there were few precedents, bar semi-fictional travel books like E. Temple Thurston's *The Flower of Gloster*.[12] Despite his publisher's attempts to dress the book like a 'country' book, with scraperboard illustrations, its purpose differed greatly from the 1930s travel books about scenic and quaint areas, like those by the rural writer and broadcaster S.P.B. Mais and those published by Batsford. It was, in a manner difficult to emulate, a record of and reflections upon a period in a life spent travelling and exploring, rather than a holiday or a travel book.

## The Journey to *Narrow Boat*

*Narrow Boat* narrated *Cressy*'s voyage from Banbury through Leicester and the Potteries to Church Minshull in Cheshire in August 1939, where Tom became employed at Rolls-Royce in Crewe for the first six weeks of the Second World War. It then covered the journey between October and December 1939 to Banbury, followed by the earliest part of a journey south towards Hungerford, in Berkshire, on the Kennet & Avon Canal.

The voyage came to halt until March 1940 while *Cressy* was icebound at Banbury, where most of the book was written. Although the book did not stress it, the second part ran at a faster pace, as he was travelling to reach new employment at Hungerford, whereas the pace of the first part of the journey (which originally intended to approach Llangollen) was very slow, with many diversions and explorations en route.

Much of the book's coverage was unrelated to the waterways themselves, as with the long descriptions of the bell-founders of Loughborough, brewing at Burton, Potteries factories and various churches. He condemned a large pub beside the canal at Blaby – the County Arms, which opened in 1938 (and has since closed) – but praised the Union Anchor Inn at North Kilworth Wharf, kept by a former boatman, Charles Woodhouse.[13] He found this 'a friendly intimate place, with an atmosphere poles apart from that of the drab and impersonal urban drink-shop'.[14] Rolt's descriptions of this pub and the canal-based Bull & Butcher at Napton, both of which soon closed, now provide a significant record that is perhaps more valuable than the themes that he sought to pursue.

The title is accurate – it covers a journey in a narrow boat and, bar the Soar to Leicester section, along waterways whose traffics were entirely carried in narrow boats. His voyage mostly traversed the waterways of the English Midlands and only the sections through Cheshire could count as the English North West.

Emphasis can be placed on scale, which probably reflected his earlier and later interest in narrow-gauge railways, and engineering, which might be seen to correspond to a human scale, in contrast to larger craft and waterways. However, these waterways involved traffic that was so limited that any long-term economic survival seemed unlikely.

His coverage of Leicester, on the Soar, illustrates much of his perspective, like the following contrast between the city by road and water:

> Broad squares and pretentious public buildings proclaim the city's commercial prosperity by road, but the water-borne traveller sees a very different picture. This is no less than the ugliness and squalor which underlie the superficial pomp and circumstance of all great cities. We saw the reeking gas-works, mountainous refuse dumps, the power-station with its gigantic steam-capped cooling towers, great mills pulsating with machinery rising sheer from the water's edge and, above all, the countless mean streets where dwelt the servants of these monsters.[15]

He did not comment that gasworks provided a surviving local traffic among those that kept the Leicester line open or that elsewhere the movement of refuse, coal to power stations and mills, and even domestic fuel to merchants who supplied some of the 'mean streets', formed staple traffics in narrow boats on many Midlands waterways. Only the canal-side location of such facilities ensured the development and survival of waterborne traffic.

And yet, in depicting navigations as occupying rarely seen spaces behind walls, in Banbury as in Leicester, he stressed their appeal to the uncovering of little-known worlds. These worlds would be distant from the tourist places in which interest was fostered by rail and, increasingly, by excursion coach travel, private motoring and, indeed, the literature of travel, but not yet by most English inland waterways outside the Broads and Thames.

Like the authors of many country books, he commented on churches, such as St Mary de Castro in Leicester, where he condemned the Victorian restorers and preferred 'the airy grace of the Early English shafts, so perfectly in harmony and balance with the graceful spring of the vaulting'[16] to the Norman work; the latter has proved to be of greater historical interest. Form, rather than historicity, was what he admired, assuming that the best and most harmonious forms were those developed by past English craftsmen.

The book provides vivid descriptions of the waterways themselves and the boats and people associated with them, as this passage about the electric battery tug at Harecastle Tunnel, which had commenced work only in 1914, reveals:

> Before long a string of horses appeared over the hill, while from the dark depths ahead a distant muttering slowly grew to a prodigious groaning and grinding sound, like that of a decrepit tramcar climbing a steep hill. At long last the tug crawled out into the sunshine, and for a few minutes the tunnel mouth was a scene of great activity, as the string of boats were detached, engines started, and horses re-attached to their respective boats. It was then our turn to be taken in tow, so we moved forward till we were beneath the overhead wire [...] For all the prodigious noise, the tug travelled even more slowly than a fully laden horse-boat, so that the dark journey seemed interminable. Lashed as she was bow and stern, there was no need to steer 'Cressy', so we sat among the sandbags on the fore-deck looking into the darkness ahead. Vivid blue sparks spluttered from the overhead conductor of the tug, and in the bar of light which streamed through the aft doors of the guard's cabin we could watch the uneven roof of dripping brickwork skimming perilously close to our cabin top.[17]

As with much in the book, this passage provides a valuable historical portrait of a process that has long ceased. In this sense, Rolt now helps to form an interpretive pivot between past users and the present pleasure users.

## Omissions and Silences

Much can be celebrated in similar vein about Rolt's descriptions of boat-building, canal-side Banbury, boats, boat families and carriers, and canal workers such as lock-keepers and lengthsmen. Some encounters seem to have blunted his acuteness of observation, producing some historical misapprehensions in the book that have been assumed to be accurate. Certain terms that seem to have derived from a misreading have passed into general understanding – 'dipper' instead of hand bowl and 'crumb drawer' instead of knife drawer, while the obtaining of water cans from a shop at Buckby, in Northamptonshire, later became the ubiquitous attribution of 'Buckby cans', rather than 'cans', as they were generally known.[18]

While the lock-keeper at Tixall confirmed Rolt's impression that the Staffordshire & Worcestershire east of Stafford was only in occasional use, a regular firebrick traffic from Brierley Hill was in fact passing. There are also some confusions about traffic – such as oil *to* rather than *from* Ellesmere Port, and narrow-boat traffic to Widnes (across the Mersey Estuary, rather than Weston Point on the Weaver Navigation).[19] Surprisingly, given the time that he spent there (q.v.), Rolt placed the major aqueduct that carries the Middlewich Branch over the Weaver above, not below, Minshull Lock.

One seemingly surprising omission is the appreciation of the actual loading and unloading of cargoes, despite the persistence of historical (and in traffic terms, outdated) approaches. Coal-loading at Bedworth and Pooley Hall is mentioned, as is a coal wharf at Banbury. There was no mention of other traffic at the latter, including gas tar and further coal wharves, of which he must have been aware through the long period when he was moored at Tooley's Yard.

The book covers much more about motive power, with approval of the declining numbers of horse-drawn boats and discussions of the past use of donkeys. Although he would later campaign for the retention and development of traffic on narrow canals, there was then perhaps less interest in the practical means of organising and handling traffic than in the people and lifestyles involved. This would tend to persist as the waterways revival movement grew in the 1950s and 1960s – the practical organisation of possible traffic would rarely be studied and measures were scarcely put into effect.[20]

While Rolt placed much emphasis upon people, oddly there is not much that is personal – his new wife, Angela, barely features and, despite his very strong opinions, few details position the author and his feelings. Despite much sympathy for boat people and their difficulties, they are depicted as archaic yet solid essential survivors in a modern world, whose impact and threat is deplored. The view – based, perhaps, on Temple Thurston's account and Rolt's own encounters with travellers at Banbury and Market Drayton – of the Romani origins of boat people was later shown to be contentious.[21]

Similarly, later researches have undermined the somewhat romantic account of the master-men, known as 'Number Ones' – owner-boatmen who had survived the competition from larger firms that had forced independent boatmen to become employees.[22] Indeed, the anonymity of many of the figures that he celebrates enables them to populate stories about the survival of 'the old proud stock of independent boatmen'[23] alongside 'the peasant and the craftsman'[24] or the countryside, without much factual corroboration. In this sense, and in the style of the book, it comes closer to the anonymity and semi-fiction of interwar country and travel books than to any factual and historical study of inland waterways.

*Narrow Boat* features much condemnation of modern housing, industry, road transport, power lines, mechanised agriculture and urbanised leisure pursuits. However, the better economic conditions (despite apparent losses in amenity) that these fostered and reflected, and the consequent improved material, if not spiritual, lives, are dismissed. While there is much Arcadian bespeaking of an older England almost seeking to return to rural dominance but with industry at craft level, little, other than regret, is offered to propose how the tide of change could be turned back, while any appreciation of the possible economic and social consequences is lacking. The second edition, produced in 1947 after the formation of the IWA, recommended IWA membership to further waterways interests, but how this would help to conserve working lives and practices was not explained.

The bespeaking of ecology is represented by descriptions of birds and fish on canals, closer to an amateur naturalist's view rather than any detailed comprehension of the ecology of canal-oriented environments. None of this should serve to undermine Rolt's appreciation of nature, craft and historic landscapes, or his portrait of the canals in decline, but it is to celebrate the form of this portrait rather than any means by which the objects of his gaze might be conserved.

Sources for *Narrow Boat* did not involve the kind of notes that a historian or travel writer might be expected to keep. The extant log of *Cressy* contains few details of evidence of canal history, although it records many observations on churches and villages en route – perhaps more conventional subjects of non-professional history at the time. There is little evidence that Rolt systematically explored canal history on the ground or in conversation, or consulted more than his very small library of books.

He 'saw interesting photographs of the Foxton inclined lift' on a visit to the nearby inn but did not follow this up with a close inspection of the inclined plane site (then only dismantled thirteen years before). At

Bedworth, on the Coventry Canal, he met a retired boatman aged 79 (and thus born in 1859/60), who 'had been over fifty years on the canal and "buried three wives different parts o' the country"'. He noted his quaint phrasing such as 'unloading them boats was billy-bally work' but recorded no details of traffic or working methods. These people must have possessed much anecdotal historical knowledge, but little of this found its way into the log or *Narrow Boat*.[25]

It is probable that he and Angela were almost the only people living on a converted narrow boat and travelling the canals of lowland England, and it would be difficult to imagine many others emulating him then, with lifestyle or vessels.[26] Some narrow boats had been converted for leisure use before 1939 with regular small-scale holiday pleasure boating over short lengths, while the Inland Cruising Association had built up a sizeable hire-boat fleet near Chester, for which Llangollen was one major destination.[27]

One predecessor was Norman Anglin of Salford, whom Rolt just missed meeting when he hired *Miranda* on the Worcestershire Avon in 1939; but, by the later period, Anglin had remained on the Avon at Pershore rather than travelling further afield.[28] However, the Inland Cruising Association fleet was laid up once war broke out, at the very time that *Cressy* arrived on the Shropshire Union Canal, so that Rolt missed encounters with canal holiday-hire boats. That *Narrow Boat* made no call for the everyday pleasure use of canals is shown by his main encounter with pleasure boats from the then recently founded Derby Motor Boat Club at Sawley, on the upper River Trent:

> A dozen or so cabin cruisers were moored head to stern along the banks, whose grass, bruised and flattened, was bestrewn with an untidy litter of paper bags, empty tins, orange peel and the embers of picnic fires. Nearly all the boats had crews aboard, of whom some were bathing, while others lolled on the decks to the accompaniment of the inevitable gramophone or radio.[29]

This reflected similar kinds of condemnations of some Thames and Broads leisure boaters at the time.[30] However, the post-war survival of the waterway track – if not the narrow boats, traffic and ways of life – could only take place with the development of more pleasure boating, alien and invasive though he clearly found it at Sawley.

## Publication, Reception and Reflections

The book, then entitled *A Painted Ship*, failed to find a publisher, and by the time it was published, Rolt had started work on two more books: *High Horse Riderless* and *Worcestershire*. His perspectives had changed slightly by December 1944, with assistance that would help his later practical campaigning. He would later see *Narrow Boat* as 'too self-consciously arcadian and picaresque. In this it was not strictly truthful [...] I find it all slightly embarrassing now, and believe its instantaneous popularity was due to the fact that it appeared at precisely the right moment.'[31]

However, the book succeeded in one of its stated purposes, 'to make a personal record of the canals and their life', which, in the words that H.J. Massingham used in his Foreword, 'were perishing under the brutal impact of industrialism'.[32] It certainly forms a *personal* record by an acute observer, even if sometimes his instincts misled him or if empathy with his perspectives is limited. These perspectives reflect a further theme – the exploration of a rural England in decline, perhaps reinforced by Massingham's influence. The book sees canals and boat people (mostly those living on narrow boats) as part of that disappearing country life. Rolt drew the stark conclusion that 'The result of my experiment has not merely proved to me the validity of an older way of life, it has left me appalled at the loss which our civilisation has sustained'.[33]

This summed up much of his life's experience to that point, coming from the apparent harmonious natural order of an older countryside to a world in which only declining traces of that apparent world could be discerned. In this sense, the book offers commentaries upon the protection of the rural that have largely gone unnoticed and have been uninfluential both in post-war preservationism and the movement to revive inland waterways.

It remains to consider the impact of *Narrow Boat* after publication and its influence. Was Rolt right in seeing the book as Arcadian, omitting so many realities? The appalling living conditions in much rural housing and employment and the grinding work and deprivations of people who lived on narrow boats were mentioned, but not emphasised. There is an element of self-caricature (but only an element) in someone from a relatively privileged background seeking to find a simpler life in one that is being lived without choice by others. Such observers can be disappointed when those who bear such roles prove to aspire to a better life with just the facilities that the observer denounces – as in Rolt's sideswipes at council housing in Alrewas or Church Minshull.

Rolt had perceived canals as 'the equivalent of some uncharted, arcadian island inhabited by simple, friendly and unselfconscious natives where I could free myself from all that I found so uncongenial in the modern world'.[34] This would repeat romantic escapist tropes, using the colonial term of 'natives', which would be largely uncontroversial then. He later saw this world as much more fragile than he had realised.

Looking back at his depiction, what has survived – in many ways astonishingly – is the track of the canals. No waterway on his route in *Narrow Boat* is unnavigable today, although parts of his route – the Leicester summit and the canal around Rugeley – were distinctly under threat in the 1960s. However, most of the same route has been transformed beyond the engineering structures. Armitage Tunnel on the Trent & Mersey was removed between 1970 and 1971, while the electric tug through Harecastle Tunnel was withdrawn in 1954 after a new ventilation system was installed.

What has almost entirely disappeared are most waterside industries and wharves, working narrow boats (and most barges), and the 'living-in' of boat people on narrow boats; the latter practice ended in the 1960s, and almost all of those concerned – workers and observers – are now deceased. Many boats have survived, now mostly owned by enthusiasts and a diminishing few by museums.

The title of the book was apt – it is about narrow boats, written from a travelling houseboat, dealing only with the lives of people upon these smaller boats of the English Midlands. There was, at the time and later, much heavier use – and greater usefulness – on the larger waterways of northern England and on the London–Birmingham route of the Grand Union Canal. His book presents a limited portrait and does not stress boats and people that were not 'gaily painted' with live-in boat people.

How accurately were the canals of the English Midlands portrayed? The final period that the book covers is the spring of 1940. Early wartime had brought disruption and decline to much waterways traffic, along with the loss of boat crews, partly to military service or wartime industries. The extent of underutilisation at a time of major pressure on railway resources was such that in 1941 the coalition government asked Frank Pick, formerly of London Transport, to report upon the potential of waterways for the war effort. His report, produced rapidly, is not a definitive account, but does suggest one realistic external view of the position.

Pick's report of May 1941, which was never published, painted a gloomy picture of both present and potential for most waterways.[35] He found that 'the carrying industry is really disorganised and that there is serious conflict of interest between various sections of it'.[36] He suggested that the canal for which *Cressy* was heading, the Kennet & Avon, should be closed as it was uneconomic, while much of the Worcester & Birmingham Canal, upon which *Cressy* would be moored for most of wartime, could be closed if its remaining traffic was transferred to road transport. Pick was trenchant about the smaller canals: 'Canals with locks capable of taking no more than a single narrow boat can not be made self supporting.'[37]

'It is a change of outlook which is the essential first step upwards,' Pick concluded, discussing 'a depressed industry'.[38] Rolt and later IWA members might well agree, but for different reasons.

Tom Rolt's converted narrow boat, *Cressy*, moored at the Market Harborough Rally on 18 August 1950. The conversion is looking its age, although it was the hull that was severely rotted. (Ian L. Wright, by courtesy of Stephen Rowson)

A renewed commercial outlook would not have rescued the elements that Rolt favoured: the small-scale independent carriers, the apparent organic ties to the rural, the traditional skills and craft in boatbuilding and canal maintenance. Pick favoured the consolidation of ownership in waterways and some carriers, the closure of less-used waterways, investment in mechanised handling facilities and commercial methods, and government support. Some of his recommendations were pursued – notably, some reorganisation of traffic arrangements and a large programme of closures of little-used waterways in 1944, including the Shropshire Union line to Llangollen, which had been Rolt's original destination. Under nationalisation from 1947, the general thrust of some of these policies would be further pursued. Rolt's world of waterways was indeed delicate and under threat, as his pessimistic second edition of 1947 suggested.

Eighty years after publication, *Narrow Boat* continues, despite Rolt's reservations, to inspire interest in waterways, although now as a historical account that can provoke many issues. It helped to inspire a movement, albeit one that was oriented around leisure, visiting the countryside rather than seeking agricultural revival. While Rolt was influenced by his new-found life on waterways, he was also pursuing ideas that would pervade his second book, and later publications. The next chapter will explore some of these ideas.

## INTERLUDE
# CHESHIRE

What is left of places highlighted by Rolt? The nature of some changes can perhaps be appreciated by an exploration of some places that the Rolts visited during their prolonged stay in Cheshire in the autumn of 1939, while Tom worked at Rolls-Royce in Crewe.

Contrary to the account in *Narrow Boat*, they first tried to manoeuvre *Cressy* into Nantwich Basin but found this inaccessible due to a bar of mud. In the 1947 edition, Rolt would bemoan the loss of Cheshire cheese that had been stored at the Basin warehouses, although even by 1939, the storage of Cheshire cheese relied on road transport. The Basin has been used since 1939 for the mooring of pleasure boats, while the surviving block of warehouses later served British Waterways' final leisure hiring operation and its successors.

The Rolts moored *Cressy* at the next bridge north of the Basin at Nantwich and visited the Star Inn at Acton. Rolt was equivocal about this pub, a building that was listed in 1967. He described it as 'an attractive example of the roadside "hedge-tavern"'.[39] 'The dark, stone-flagged interior was rendered even darker by a typical Victorian stuffiness of wallpaper, aspidistras, lace curtains and coloured lithographs,' he complained, but preferred this to 'exploitation by the builders of Tudor roadhouses and "olde-worlde" tea-barns'.

Acton, then little more than a hamlet, is now much expanded, with a small 1950s housing estate near the canal and more recent housing. The path from the canal bridge to the main road has been diverted to pass into and down Wilbraham Road. The Star closed in 2015, as did the Jolly Tar at Barbridge Junction, further north, but unlike the latter, whose site is now occupied by high-value housing, the building has been returned to something closer to its earlier appearance in a sensitive conversion to a residence, while new housing occupies the site of the car park.

In Nantwich town, which he saw as a market town already seemingly under siege from the expansion of Crewe, Rolt viewed surviving timber-framed buildings, notably Churche's Mansion in Hospital Street.

He greatly commended it 'as fine a flowering of the regional style', but regretted that it should have become a tea and antique shop.[40] This was itself preferable to the 1930 plans to dismantle the whole building and ship it to America for reconstruction, a fate prevented by the then owners. In 1947, he noted that the building 'stands empty and fast falling to complete ruin'. This did not prove to be its fate; it later became an antiques shop and most recently a restaurant. Rolt would, however, no doubt be as uneasy about this as he was about the leisure industry and canals, coming to see this as a lesser evil. But the shops in the town no longer cater for 'the countrymen of Cheshire'. Those that survive serve a suburban town, while the weekly cheese markets in Nantwich ceased many years ago.

For six weeks from the beginning of war in September 1939, Rolt moored *Cressy* at Church Minshull Wharf so that he could travel to the Rolls-Royce factory in Crewe by motorcycle. The Wharf is situated above the Weaver Valley at a short distance from the village, reached via Cross Lane. The wharf had ceased to be used in the 1930s, but the wharfingers house, now named Minshull Wharf, remains, with a shepherd's hut in the back garden. The stone bank of the wharf edge is still in place, as are the stone steps and gate leading up to the bridge that carries Cross Lane, which Tom must have used daily. The stone wharf may well have been one of the few places where a 70ft boat could moor. Walking down this lane, the narrow bridge here, listed in 1984, provides the only road crossing of the Weaver between Winsford and Nantwich.

In 1945, the Weaver Navigation obtained powers to extend its navigation for large vessels upstream from Winsford to a large new depot outside Nantwich. This was never built, but if it had been, the impact on Church Minshull would have been considerable, with a large new road bridge, extensive changes to levels and probably a very large aqueduct to carry the Middlewich Branch over the new river navigation. Had this proceeded in the post-war period, it would have presented the IWA with a conservation dilemma. It had been founded to promote the use and development of waterways, including new waterways, but the latter would have had a considerable impact on places like Church Minshull. Rolt never favoured the enlargement of narrow canals, feeling that it would suffice to restore them to their original dimensions.[41] Without the planning system, which incorporated controls to protect historic structures, the bridge would have been replaced, yet Rolt was opposed to the system when it was made universal after 1947.

Beyond the minor highway leading to the Wharf and on to Crewe, Church Minshull now stands on the busy B5074 road between Nantwich and Winsford. Rolt regretted that this road, which had been surfaced in fine cobbles, had only recently been covered in tarmac, 'it was this funereal road alone that brought the twentieth century to Church Minshull, for at night, when the street was dark and still, the village was ageless'.[42] This would certainly not be the case now – like many Cheshire villages, most properties have been acquired by commuters and retirees, with steady streams of traffic.

Many historic buildings have survived, however. Rolt would be able to recognise much of the centre of the village, if not the more recent housing on its outskirts. The roadside front of the Badger (it was formerly the Brock or Brook Arms and the Old Badger for a time in mid-century) appears little changed from the exterior. However, there have been major extensions, including one into a former coach house, which had been a garage with roadside petrol pumps. After a period of closure in the 2000s, since 2011 this has become a popular gastro-pub, serving 'British cuisine'. Rolt's 'gleaming and spotless taproom' seems to be that which has 'bar' on the glazed window in its door; this looks largely unchanged, although most of the rest of the building has been extended.

The social purpose of this pub, which was for local men (but not women) to drink after work, has diminished, although few people now work in the village. Services are now only held part-time at the adjacent Anglican church, St Bartholomew's, where Rolt had examined the parish registers, describing the pub and church as the core of rural life, as perhaps they both were in the past.[43]

Opposite, as the road turns towards Winsford, is the site of the post office, where Rolt noted another centre, with the postmistress, 'white-haired and benevolent, as such a personage should be'. The building is listed with a blue plaque, and still stands, but the post office closed in 1953.[44]

Rolt seems to have been unaware of the social relations in the village, which was owned by local landowners the Brooke family until a series of sales between 1920 and 1923. The postmistress, Amelia Brereton (1870–1960) later became a councillor, representing Liberal interests against Tory local landowners.

After the post office was relocated to the western end of the village, this too closed in the 1990s. The complex opposite, partly converted farm buildings and partly recent construction, might well have drawn his ire and despair – a modern residential development, clearly aiming to attract commuters and retirees, who must now form most of the village residents.

Around the corner is the site of Minshull Mill, upon which Rolt focused extensive discussion as this was a surviving watermill, using the fall of the River Weaver, which also generated electricity for local use. Again, the building remains, now converted into apartments, but as electricity supplies were interrupted when the Weaver was in flood, a more regular public supply was secured in the late 1950s and the surrender of the water rights was sought in 1955. Nearby is the site of the Smithy, where Rolt commended a working blacksmith, Mr Eggerton, 'wheelwright as well as smith', with horse-drawn farmers' gigs and carts brought for renovation.

It is, again, 'Rolt country': while buildings have survived due to becoming listed, the trades that made Minshull a rural community have disappeared, one by one. It is what he feared and anticipated, and the apparent stability, even as war broke out, that he described did not last long.

Judging by his comments in *Narrow Boat*, Tom Rolt would not be impressed with the interior decoration of the Star in Acton, shown here in 2005.

The Star in 2005, with the village of Acton beyond. At the time of writing, the former Star has been converted to a house closer to its historic appearance than this half-timbered front elevation.

Looking east, Church Minshull Wharf on the Shropshire Union Canal is on the left. *Cressy* was moored here for six weeks in the autumn of 1939.

The former Smithy and Old Post Office, flanking the corner of Cross Lane in Church Minshull.

The Old Post Office in Church Minshull. The road ahead crosses the River Weaver.

The Badger pub in Church Minshull, looking west.

# 3

# INFLUENCES AND INVOLVEMENTS: MASSINGHAM, ORGANICISM AND LITERATURE

Some of the deep regret and sense of decline that features in *Narrow Boat* derived from Rolt's developing reading and thinking, which continued during the long years from 1941 to 1946, while *Cressy* was moored at Tardebigge in Worcestershire. Before and after the book was completed, Rolt was influenced by reading and correspondence that reinforced some of his earlier perceptions – some of these from his interpretations of childhood encounters. He wrote later that 'for me, doctrine has always been accepted or rejected by the touchstone of personal experience and not the other way round'.[1] While at Tardebigge, his extensive reading led him to write for various magazines. He was prolific, and perhaps precocious, with magazines: his main theme became 'the baneful effect of modern technology and economics on world ecology, particularly as applied to agriculture'.[2]

This eventually resulted in his second book, *High Horse Riderless*, which reflected his contact with political ideas that embodied an organicist approach to conservation. The politics seems largely to have been vicarious, through his contact with Harold John Massingham (1888–1952), who was a leading organicist writer connected with various controversial individuals on the fringe Right of British politics, and with Massingham's own involvements.[3] This chapter sets out to contextualise this development and to explain Rolt's role. Considering the growing views for varied forms of rural conservation, I then attempt to set the inspiration for Rolt's first two books within the literature – the literary landscapes – of the interwar years. This may help to explain the contrasting reception of each book.

## Harold John Massingham and Organicism

The politics of organicism, seeking large-scale social transformation, was derived partly from Massingham, who greatly encouraged Rolt in the 1940s. Massingham, a generation older than Rolt, was a prolific and popular writer on countryside and conservation, publishing numerous books and magazine articles.

Massingham's views, increasingly trenchantly expressed, had evolved since 1918. The son of a Liberal journalist, he had met the rural writer W.H. Hudson (1841–1922) four years before his death, and became involved with nature conservation in 1919, with the Plumage Group, which aimed to ban the use of birds' feathers. His first wife, Gertrude Speedwell Massingham, was a Labour Parliamentary candidate in 1929 and 1931, while Massingham was then influenced by Guild Socialism. After he had divorced and moved to Long Crendon, in rural Buckinghamshire, in 1932, he wrote numerous countryside books with a progressively conservative outlook.

The Foreword to an anthology of Massingham's work asserted that 'H.J. Massingham was plagued throughout his career by this word, peasantry' and considered how this term had changed in use. Both he and Edward Hyams (who came from an explicitly socialist background but who co-wrote Massingham's last book) advocated self-sufficiency, especially in agriculture, and suggested that 'this could be achieved without any radical innovations, simply by restoring the traditional English economy based on village communities of cultivators and craftsmen'.[4] Rolt's views would tend to partly reflect this idea of a simple revival.

From the late 1930s, Massingham seems to have turned from a sentimental interest in the rural, agriculture and villages, as well as a deep commitment to nature conservation, towards political allies in a movement attempting a shift towards 'restoring the traditional English economy'. Organicism would later be associated with the 'Green' movement, which today would tend to imply support for broadly Left/Liberal positions, as with many (but not all) Green parties in the world. At the end of the 1930s and into the 1940s, however, concern for environmental issues propelled adherents towards highly conservative, if not fringe or far right-wing positions. As the far Right is now in its ascendancy in several European countries, it is worth contrasting different kinds of right-wing politics.

In Rolt and Massingham's time, a significant current within the extreme Right was fascism: this favoured a strong corporate state, a militarisation of society, authoritarian centralised control, and ruthless infrastructural, urban and rural development. The latter might feature variations, as with the Nazi 'blood and soil' tendency. Rolt (and Massingham) would be explicitly opposed to much of this, favouring a rural/agricultural revival, which would restore the squire, parson and peasantry to a pre-industrial position within what they saw to be part of a natural order, as will be discussed later. Both would feel that the forces of industrialised agriculture should be limited, while an agriculture rooted in traditional craft would be encouraged.

This corresponded to a deeply conservative viewpoint and possibly, to achieve its implementation, this might have necessitated a quiet form of rural neo-fascism involving a different base from the militarised urban version. The historian Arthur Bryant had expressed something similar in 1929:

> From the plain man has been taken away the home smoke rising in the valley, the call of the hours from the belfry, the field of rooks and elms [...] And the spirit of the past – that sweet and lovely breath of Conservatism – can scarcely touch him. It is for modern Toryism to recreate a world of genial social hours and loved places, upon which the conservative heart of Everyman can cast anchor.[5]

It is doubtful that either Massingham or Rolt would dissent much from this conflation of this vision of a rural idyll with a 'traditional' social order – with the past and with the beauty of the countryside – although there was a more apocalyptic view that saw this as not just desirable but essential if 'civilisation' was to be restored.[6]

One feature of the organicist political Right is that it had little public support. It comprised isolated elites of writers, scientists and public figures, but ones that did not attract or develop any sort of mass movement, as Oswald Mosley's Fascists, and later the Greens, would. Rolt's associations with Massingham and, through him, other figures, would place him firmly on the political Right – a question that needs to be addressed.

The fringe Right involved many differences in emphasis. Massingham was associated with various organicist British figures, who included pro-Germans, pro-Nazis and anti-Semites. Some had been attracted by admiration for the Nazi 'blood and soil' approach to agricultural policy, while others, like Rolf Gardiner, were attracted to the idea of restoring a feudal order headed by an absolute monarchy.[7] Massingham had co-founded the Kinship in Husbandry in 1941 with Gardiner and Viscount Lymington (Gerard Wallop, a former Conservative MP).[8] Confined to twelve members, this would also include Arthur Bryant. However, Massingham seems to have pulled away from them on the grounds that they were pro-German.[9] Nevertheless, his *The English Countryman*, published in Autumn 1942, is dedicated to 'my fellow-members of the Kinship of Husbandry'.[10]

Rolt does not seem to have joined any of these organisations, but his second book, *High Horse Riderless*, reflected some of the ideas expounded by Massingham.[11] He had subscribed to the *New English Weekly*, at Massingham's suggestion, and in February 1944 Massingham had introduced him to the Kinship in Husbandry, despite its domination by Rolf Gardiner, whom he described as 'a menace'.

Rolt expressed trenchant views, as will be discussed later, but it may be that his desire to revive agricultural society drove him towards the organicist Right, rather than an organicist position leading him to support policies to revive agriculture. He would later be involved in practical politics of lesser scope, dealing with the IWA, and his support for wider social change seemed then to be rhetorical. There was no question that he was deeply suspicious, even of the social democrats, who would form the post-war Labour government and who had influenced wartime policy. Numerous exchanges with Robert Aickman, co-founder of the IWA, make this clear; for instance, Francis Klingender, who had written about popular arts in the Industrial Revolution, was simply dismissed as 'very Left', while there would be suspicion of Charles Hadfield, who was very much on the right wing of the Labour Party.[12] While averse to anything that advocated State control or collectivism, he laid stress on individual freedom, as will be discussed later. His was a complex and not always consistent position, which would change from the 1950s onwards, and so a straight description as 'right-wing' could mislead.

In 1945, in *The Natural Order*, Massingham would advocate a 'Return to husbandry', and as *High Horse Riderless* would reflect, 'We shall not, in fact, begin to understand the meaning of husbandry unless we relate it to the first principles of the natural law, which is an earthly manifestation of the eternal law'.

## Massingham and *High Horse Riderless*

A surviving file of letters from Massingham enables some discussion of Rolt's evolving views towards *High Horse Riderless*.[13] The correspondence was often intense and began with a 'fan' letter to Massingham, seemingly on 29 September 1939, when Rolt was living on *Cressy* at Church Minshull in Cheshire. Massingham agreed to examine a manuscript of Rolt's, which was probably his novel *Strange Vista*.

In a letter of 22 June 1943, Massingham waxed eloquent about the manuscript of *Narrow Boat*. He wrote, 'I was so enchanted with it that I felt I could not rest until I saw it on its way to the printers.' 'You will guess that I was in complete sympathy with its point of view.' By the time *High Horse Riderless* was drafted, Massingham seems to have further influenced Rolt in both the form of his writing and in the politics that this involved.

Rolt had sometimes used apocalyptic language in *Narrow Boat*; for instance, he described a combine harvester in Leicestershire as 'man having declared war on Nature and brought up his heavy artillery'.[14] He now developed similar, if stronger, language in *High Horse Riderless*. This was named after a phrase in W.B. Yeats' poem 'Coole Park and Ballylee, 1931', which, in a complex manner, regretted the passing of an earlier era. It is notable that when Rolt wrote in wartime, it 'was a pessimistic book written in an optimistic mood. When I wrote it I had felt so certain that there would be a change of heart after the war.'[15]

His book was presumably a contribution to that change of heart. It opened with some autobiographical observations. Through his work in industry, he discovered that 'individual skill and responsibility was a liability which was being eliminated from industry with bewildering speed'.[16] Later, he had endorsed the 'prospect of a return to the natural world which now seemed to me [...] the only reality'.[17]

He stressed again that the book was based largely on experience: 'life has been too short to be both practical man and scholar.'[18]

He developed a set of ideas that were coherent in their own terms, although dubious in many respects. He felt that medieval societies had corresponded to a natural order, governed by a moral law: the universal *jus naturale*. The secularisation of the Christian Church (both he and Massingham were of Anglican heritage) had changed attitudes to the natural world and made it more into a force to be conquered than one whose laws had to be followed. This had led to mechanisation and the 'acquisitive society', which involved wastage and the exhaustion of resources.

Rolt would go further in his novel *Winterstoke* (1954), in which he praised a stable order where monasteries, guilds and wealth were 'governed or restrained by that sensibility and good taste which is only acquired by time and experience'.[19] In the novel's final sentences he stated: 'After four hundred years the Cistercians have come back to Winterstoke and confront another wilderness.'[20] Rolt implied that industrial civilisation presented a new wilderness that this monastic order (among others) would have to clear and make perfect. The Cistercian emphasis on the worth of manual work – for the glory of God, in their doctrine – stressed that any future would have to be based ultimately on manual labour, albeit applied to ends that humanise the environment and corresponding social order.

Rolt saw a first step as being the application of the principles of natural law to education. A more enlightened education would inform a generation that would then rigorously oppose the stultifying conditions of life and work. His chapter, entitled 'Education for Freedom' – a freedom that was little defined – then presented a curious mix of intriguing ideas with a lack of programme as to how any part could be implemented. His opening points included a startling assertion that education's 'cardinal sin has been its tendency to inculcate the idea of ambition as a virtue instead of the perversion which it is'.[21] He referred to this as a 'careerist incentive'.

While he acknowledged that political change would be difficult, those who supported 'the value of self-expression [...] can put on the armour of their particular ability to defend the coming generation by ensuring that as many as possible shall receive such an education that they will not lightly submit to state tyranny'.[22] Commenting on the content of education, he condemned the 'wilful arrogance of scientific materialism which ignores the past, and on the other by the sentimental nostalgia which accepts the past uncritically'.[23] He called for study of the 'common man through the ages'.

Rolt praised those who had secured the preservation of past objects, like tools, 'telling of their wise village organisation of common field, court-leet and manorial custom, of how they ploughed, sowed and reaped, and celebrated that fruitful partnership between labour and natural bounty in gracious traditional ceremony'. This harmonious portrait had been destroyed, he asserted, by the parliamentary enclosure process, whereby common field systems were progressively abolished in favour of enclosed fields and new roads in the interests of agricultural efficiency.

There is much to question in this, despite some interesting ideas that could be explored further. Education systems and policies are very complex, with change hard to implement, and there seemed little appreciation of how this might all develop, who might favour and secure change, and how this future education might be delivered. The last contained a sideswipe, which perhaps reflected Rolt's own negative experiences of schooling: 'All too often the task of education has been left at the mercy of time-servers attracted to the profession by the sheltered life, long holidays and eventual pension.' He saw hope in 'the artist', although how future artists would be supported and be able to practise their arts as well as teach was left unclear. He advocated supplementing permanent school staff with contributions from all, imparting 'their experience for the benefit of coming generations'. This laudable principle does beg the question as to how all of this could be done.

It seems that if a national system of education on Rolt's principles could be established, and:

> ... we were able to slowly and painfully restore the associations and spiritual values we have lost, and with them the conception of true freedom, what then follows? How would a generation so assured of its place in time and space confront the hostile present of our existing social structure?

Despite his confidence that it would set about rebuilding those structures, one wonders at the intergenerational conflicts and how these might be resolved. He acknowledged that this might prove to be long term: 'We can sow the seed of a new and better civilization tomorrow, if we wish, though it may be two hundred years before it come to flower.'[24]

It is clear that this is a mixture of conservative and radical principles, but far distant from the kind of rigid authoritarianism that fascist approaches to education would take.

The consequences would include a self-sufficient society, making the fullest use of human ability and natural resources. The basis for self-sufficiency should lie in prosperous agricultural communities, making industry's prime purpose the support of agriculture. His conclusion was 'That a self-sufficient society would, through the realism of its agricultural basis, re-affirm the validity of a natural order, recover the spiritual values of religion, and perceive with reason the necessity of moral law'.[25]

Written in wartime, this provided many themes that would be echoed in the later environmental and Green movements: self-sufficiency, wherever possible, and the devolution of power to local communities; the possible exhaustion of natural resources; the need for suitable education; and critiques of industrial and urban development and consumerism, with new emphases on sustainable agriculture. Not discussed in *High Horse Riderless* was the opposition to the use of nuclear weapons (and later nuclear energy) – or, indeed, conventional warfare and defence – but Rolt would later record how his optimism turned to despair at news of the atomic bombings of Hiroshima and Nagasaki in Japan and revelations about the Nazi death camps. Few would then support the views that he expressed, but these were prescient – so much so that Green Books reprinted his book in 1988. He recorded that *High Horse Riderless* sold few copies, much like his later *Winterstoke*, unlike the major success of *Narrow Boat*.

Massingham commented in detail on the draft of *High Horse Riderless* on 14 November 1943, suggesting further reading. On two undated sheets from 1943, Massingham assessed the manuscript: 'As it stands, I wouldn't send it to a publisher.' Rolt had given 'the wrong reasons for the downfall of mediaeval Christianity – yours is the average 19th century view', and more historical research was required with more concrete examples. Massingham also insisted that more referencing was needed and, although some ideas might be original to Rolt, others might well have formulated similar ideas and should be cited as authorities.

Rolt does not seem to have followed much of this advice. However, Massingham was fulsome in his praise, knowing 'of none, not one, except yourself, who has a coherent, constructive, properly pondered philosophy of life to expound. And I call that achievement extraordinary. Especially as it is the right one!'

How realistic was the driving vision of *High Horse Riderless*? And how desirable would change have proved, had the book inspired it, in the post-war world? Present politics in the UK (and indeed elsewhere) include so many variations that it is impossible to assume some sort of consensus or an objective position from which this writer can pronounce. For what it is worth, I have major sympathies with a Green position, but not one that would attempt any return to some sort of golden socialecological age. I too would call for a balance between humans and nature, but not from any spiritual viewpoint, while any viable future will require the development and application of appropriate new technologies.

To fully assess Rolt's ideas in *High Horse Riderless* would require a detailed examination of the whole aim and methods of environmental politics since the 1940s. One area in which there would be later echoes was his insistence, reflecting his engineering background, that technologies should be applied and not abandoned – but intermediate technologies, similar to those advocated by E.F. Schumacher's *Small is Beautiful*. To encourage greater use of Midlands canals, for instance, he would advocate the improvement of handling methods and the enlargement of narrow-boat cabins. There was potential conflict there between the conservation of the historic and the conservation of resources.

The manifesto set out in *High Horse Riderless* left it entirely unclear who would be in a position to take note of it and identify forces that could take the favoured changes forward. The book took the form of many such publications: to set out what was desirable in the hope that an unidentified readership would be convinced. The prospects of influencing those who held power – in states and owners of capital – were neither identified nor realistic. Even the then Churchill coalition and the Labour opposition that would soon gain office (if not power) were both concerned with post-war reconstruction. That meant economic development and the general improvement of living standards, rather than any turn towards a vision of self-sufficiency that perhaps owed too much to sentimental views of the countryside or to landowning backwoodsmen, like Rolf Gardiner and Viscount Lymington, who were very much on the fringes of upper- and middle-class society. The reduced food production that organic practices in agriculture might involve – as opposed to more chemical, industrialised methods – would have had major consequences, even in an economy in which much food was rationed.[26]

How desirable were the changes that Rolt and other organicists advocated? Those who had suffered from wartime shortages or experienced evacuation to rural areas might find it hard to support a movement that could make such privations permanent. Similarly, the implied restrictions on tourism and holidays, much suppressed in wartime, would go against the wishes and needs of many. Indeed, it would be tourism and holiday boating that would enable many canals to stay open. The appeal of small rural communities, which provided havens and retreats for visitors from urban areas, might well be limited for many who lived in those areas, whether these were under the rule of paternalistic landowners or in areas where property ownership was more diverse. Similarly, the appeal of monastic orders like the Cistercians might be dubious for those who might be expected to participate and accept the extreme controls over personal life that this would involve.

One possibility for a more sustainable future would have involved planning, with tight controls and public investment in rural revival, but Rolt explicitly disavowed this. He viewed it as totalitarian, in a way that he would not view any order dominated by controlling landowners or corporate capital, however benevolent in intent. Matless' distinction between organicists and 'planner-preservationists', often seen as in concert, is very helpful here.[27] 'Planner-preservationists' would pursue various policies, including the containment of urban settlements, support for viable rural communities, the promotion of protection for scenic areas, new towns in appropriate locations, the protection and conservation of historic structures, and, to a lesser explicit extent, nature conservation. This would reconcile modernisation with conservation.

One aim of planning was to avoid the wasteful misuse of land in ribbon development and urban fringes, to support agriculture (admittedly, assuming that it could ensure a satisfactory countryside in aesthetic, economic and amenity terms) and to plan the suitable development of viable settlements.

In some ways, this should not have been so distant from the ideas that Rolt and others held. But ideas like the restoration of the peasantry, the conservation of crafts as a basis for a consensual society and the promotion of self-sufficiency amid a rigid hierarchical society on medieval lines proved far apart, while the whole analysis of how and where the Church had given way to secularism and the view of the Industrial Revolution as aberrant could not chime with the modernisers of the 'planner-preservationists'.

## 'Middlebrow' and the Reading Public

To appreciate landscape, to be inspired by it and to consider the underlying processes involved may lead some to write about it in diverse ways. Rolt, like Raymond Williams, sought to write fiction at first, but then mostly wrote non-fiction. He did not, however, seem to have been inspired by experiences in formal school education, unlike Williams, who would go on to teach in adult education and later became a professor at Cambridge University. Rolt's schooling had set him against the arts, like ballet, opera, classical music and even Shakespeare – although Massingham and Rolt's earlier companion 'Anna' swayed his opinion of Shakespeare, of whom he came to approve.

It would seem dubious to trace various people who inspired Rolt, like W.B. Yeats, and then to place him as some sort of successor. The popularised term 'middlebrow' would seem to apply to much of his work.[28] It has been used in a pejorative sense as a term of condemnation, but it can form a more neutral description of popular works, written with serious intent to inform and educate a more general public than the kind of elites that 'highbrow' literature, art or music would be expected to serve.[29] This would usually be a middle-class audience at whom radio and some parts of film cinema would be aimed. Rolt had noted that the term 'highbrow' was used to accuse the artist – 'the only custodian of values which our civilisation has discarded'.[30] The same readership from the middle class would tend to share some of Rolt's conservatism, although not always his aversion to suburban development and living.

Rolt would write for a middle-class public, one that was oriented around his particular concerns, but not necessarily a waterways or rail-enthusiasts audience. What literature had shaped his potential audience, who might find *Narrow Boat* and later publications of interest?

He recalled that there were few books at Cusop – the only children's book he recalled (appropriately enough, perhaps) was *The Secret Garden*. An early influence was Samuel Smiles' *Lives of the Engineers*, which he acquired in 1932, at much the same time as he became interested in the poetry of W.B. Yeats.[31]

One of Rolt's favourite poems by Yeats was 'Nineteen Hundred and Nineteen', reflecting on the Irish Civil War. Yeats had written in pessimistic vein: 'Now days are dragon-driven, the nightmare Rides upon sleep.'

Rolt admired Aldous Huxley, who is little read today. His *Brave New World* (1932) depicted a future dystopia, in which mechanisation and genetic engineering have taken over. Rolt commented that this 'seemed to me to portray the kind of future to which I felt that industrial man was heading'.[32]

Huxley was popular and erudite, but disillusioned. As Harry Blamires put it, 'His registration of post-war society seems to mock an exhausted culture whose religion and morality are spent, whose ideals have evaporated, and whose notions of romantic love have faded into animal promiscuity.'[33] Rolt acquired *Ends and Means*, a book of essays, just after its publication in 1937, but it is unclear whether he would agree with Huxley's views on 'reform'.

After 1932, Rolt wrote a novel, *Strange Vista*, which covered a developing locality in rural England between the mid-eighteenth century and 1954, his deemed future. Following a breakdown of civilisation, his main characters found refuge on the Welsh border in 'a life of Arcadian simplicity', evoking William Morris. He was perhaps influenced by the interwar depression, which had terminated his apprentice employment in the Potteries.

## Regional and Rural Novels

Fiction with a regional setting was particularly important in the interwar period. Rolt may have read Mary Webb's novels of rural Shropshire, set in ambiguous (but distant) historical periods. After her death in 1927, her work was commended in 1928 by Prime Minister Stanley Baldwin, and became very popular in the 1930s. By 1931, there was a book entitled *The Shropshire of Mary Webb*, and by the mid-1930s, charabanc excursions were being run to this area, south of Shrewsbury.[34]

Later, Rolt recalled that Francis Brett Young was the only fiction writer whom he enjoyed reading: 'His descriptions of landscape, of weather and season are masterly evocations.' An early example of Young's work, *The Young Physician* (written between 1917 and 1919), is set in 'North Bromwich' (Birmingham). The closing page quotes Thomas Traherne: 'You shall never enjoy this world aright until the sea itself flows in your veins, till you are clothed with the heavens and crowned with the stars.'[35] Set in 1895, with the water scheme to supply Birmingham under way, Book 2 begins with an epigram from Traherne: 'and so with much ado I was corrupted, and made to learn the dirty devices of this world.'[36] Traherne and Henry Vaughan, who knew the same border areas as Rolt and were perhaps inspired by similar landscapes, provided inspiration.

In Young's book, corruption seems to apply to the urban world, with hills converted to tramway gradients and rivers channelled into brick culverts. The physical despoiling of the 'North Bromwich' landscape has been matched by spiritual loss:

> It is not a place from which men have wilfully cast out beauty so much as one from which beauty has vanished in spite of man's pitiful

aspirations to preserve it. Indeed, its citizens are objects rather for pity than for reproach, and would be astonished to receive either, for many of them are wealthy, and from their childhood, knowing no better, have believed that wealth is a justification and an apology for every mortal sin from ugliness to original sin.[37]

On the other hand, from Pen Beacon, 'with a choice of prospects, one may turn from the dreamy plain of Severn and the cloudy splendours of Silurian hills [...] and perhaps in that remoter air you may realise the city's true significance as a phenomenon of unconquered if not inevitable disease'.[38] Rolt would follow much of this imagery: the urban and industrial deemed to be corrupt, the rural deemed to be pure, innocent and under attack, and a lost and declining natural and cultural heritage. It is noteworthy that, as in Webb's novels, the setting is in a distant past. It is hard to know when 'The Fall' into a corrupt world took place.

Williams' comments on Young are useful here: 'The loved places are the "unspoiled" places, and no group agrees with this more readily than those who lived in the "spoiled".'[39] It is, in the end, an urban vision idealising the rural.

## Interwar and the Rural Idyll in Literature

The 1930s saw an increase in the number of regional and rural novels, making a contrast with the modern England that was often portrayed by London-centred writers. The latter was perhaps exemplified in J.B. Priestley's *English Journey* (1934), which posited three Englands, one of which was the older industrial England contrasted with the England of cathedrals and market towns, both set against a modernising landscape that included bypass roads and large dance halls. Kristin Bluemel has stressed the contradiction in the repudiation of the latter, which Rolt certainly recorded in his depiction of the public house at Leicester: 'Mass culture erases historically significant differences of region – of place and dialect and custom [...] On the other hand, its roads and vehicles provide broad access to English and British regions at the very moment its popular mass print culture publications fuelled the public affection for the countryside.'[40]

What of the 'reading public' at the time? Much of this can be traced to the middle-class 'popular' rather than the 'highbrow', and if Rolt could be described as an intellectual, this was, as he often stressed, through experience and cogitation rather than training, credentials or admission to an elite. Popular books of the 1920s and 1930s would be those most in circulation in public and circulating libraries.

One theme that ran through many, which can be traced back to the 1914 war (and the view of the Edwardian period), is that of the rural idyll, often with the countryside and 'traditional' agriculture under threat. In non-fiction, this included the books by, or edited by, Clough Williams-Ellis, which underlay the formation of the Council for the Preservation of Rural England (CPRE) in 1926 (and Council for the Preservation of Rural Wales (CPRW) in 1928). There were also travel books, notably those by S.P.B. Mais, which were partly based on radio broadcasts and which Rolt may have heard about – he had acquired a crystal wireless set at school in the early 1920s. *This Unknown Island* (1932), for instance, was based on BBC talks 'for the purpose of stimulating in listeners a desire to explore and rediscover their own island'. If 'listeners did not want to rush off at once and explore the district through which I had just rushed,' Mais wrote, 'I had failed entirely in my purpose'.[41] He focused on lesser-known places like the 'Mary Webb Country', including the Stiperstones and Long Mynd Hills in south Shropshire.

Although they were written in the 1870s, it was not until 1938–40 that the diaries of Francis Kilvert were published. Kilvert had been curate at Clyro, near Hay-on-Wye, and portrayed a world that Rolt felt had changed little subsequently (although he had left the area in 1921). Once again, the endangered heritage of the past was at a distance. The diaries reveal much about poverty and contagious disease in rural areas. As Ronald Blythe put it, 'We have a tendency to read Kilvert's Diary as an idyll, but it is in many ways a shocking and unsparing description of the Victorian village.'[42] Kilvert's response was pity rather than outrage and a call for reform. There may be some parallels with the later Rolt, when Blythe again characterised Kilvert as 'the determined intruder, an adventurer along the footpaths, a watcher of lives which fascinate him but which he instinctively realises are beyond his jurisdiction and outside his experience'.[43]

The themes of a lost organic rural world were continued in Richard Llewellyn's *How Green Was My Valley*, which would be made into a sentimental Hollywood film; this was one of the most popular books of 1939.[44] Rolt had not yet read this when Massingham commended it in 1943 – 'It's soaked in poetry and locality. I was immensely struck with it.'[45]

The audience for *Narrow Boat* may well have been similar to that for Llewellyn's book. This largely depicted a late-Victorian period from which relations between nature and humans and employers and workers in a south Wales mining community had deteriorated. Large-scale coal mining had destroyed the earlier organic relationship between people and their environment. More sinisterly, the 'organic' community had been damaged, not just by the changes in mine ownership but by the contamination by 'half-breeds', one of whom is 'executed' by representatives of the 'community' after allegations of child abuse.

Its evocations of a distant past, a seeming golden age, also had resonances. There was, it seems, always a period in the past when all had been much more wholesome, and Rolt's first book seemed to suggest that one could make a return to a past period – in the imagination if not in reality.

On the borders between memoir and fiction, Flora Thompson's *Lark Rise* (1939) was the first of a trilogy, published together as *Lark Rise to Candleford* in 1945. This similarly evoked traces of a lost world in a rural Oxfordshire hamlet affected by enclosure.[46] Rolt must have been aware of Massingham's 1944 introduction to Thompson's combined volumes, wherein he denounced 'the utter ruin of a closely built organic society with a richly interwoven and traditional culture that had defied every change, every aggression, except the one that established the modern world'.[47]

All these works were popular and, regardless of whether Rolt read and was influenced by them, will have been of interest to much of the potential readership for *Narrow Boat*.

## Some Conclusions

Several interwar regional and rural novels (and other forms, like travel books and film) provided an image of the rural that readers may have shared with Rolt. Raymond Williams' *The Country and the City* characterised the country, on which 'has gathered the idea of a natural way of life: of peace, innocence and simple virtue'. Unlike Rolt, he saw positive coverage of the city, with 'the idea of an achieved centre: of learning, communication, light'. He also noted 'powerful hostile associations' – 'limitation' in the country, 'ambition' in the city.

As Williams asserted, 'The real history, throughout, has been astonishingly varied'.[48] His book dealt with persistent images of both but noted that despite urbanisation, much literature remained rural.

A contrasting view was that 'The village was the bedrock of English rural society with the church at its centre, surrounded by farms, cottages and fields. This for Massingham was the rural trinity of God, Man, Earth, the simple basis of his faith [...] For Massingham a return to nature meant a return to God.'[49] How far this was illusory is another matter.

Rolt had sought to go beyond the celebration of the rural to consider much wider issues, no less than the whole path of development in (at least) Britain. He might have hoped that some might reflect and act upon *High Horse Riderless* but he had no such intention for *Narrow Boat*. Yet it was *Narrow Boat* that inspired public interest and helped to found a movement.

INTERLUDE
# CHESTER

Although the Rolts left suburban Chester when Tom was 4, he continued to spend Christmas holidays there with his uncle, Dr George Taylor, until at least 1921. In 1971, he continued to assert that 'Chester is a Roman city', and in childhood he stood on the walls overlooking the Roodee racecourse and Dee, imagining himself as a Roman legionary watching for invaders from Wales.[50] This section of the walls was, however, a twelfth-century extension, made into a leisure walk in the early 1700s when their defensive function was no longer required.

Rolt seems to have developed an acute feeling for historical places from a very early age. In 1971, he wrote, 'The Edwardian world into which I was born appears as remote and fantastic as some half-remembered dream'.[51] An early feeling for one version of the past is illustrated by a childhood memory of one Christmas Eve in Chester:

> Our way took us down Watergate Street. This was a narrow cobbled road flanked by even narrower pavements and shadowed by the over-sailing timbered gables of the houses. Because both street and pavements were thickly carpeted by new-fallen snow, we climbed the stone steps to the shelter of the row. Unlike the more frequented rows with their gay shop fronts, Watergate Row was a dim, mysterious place after dark, lit only by infrequent flickering gas lamps. It was inhabited — as it doubtless had been since the Middle Ages — by small craftsmen, coopers, tinsmiths and the like. The windows of their workshops were shuttered now, though some showed chinks of light and there were sounds of unknown activity within. Occasionally the mouth of a narrow alley, dark as midnight and leading who knows where, opened up between them [...] Attempting to formulate at this distance in time I would say that what I perceived then was an embodiment of the continuing life of an ancient city, labyrinthine, dark, mysterious yet not sinister but intensely human.[52]

An antiquarian might stress that this part of Chester was, in fact, largely a Victorian reconstruction rather than a medieval survival, and most of these shops were then, as they would be later, antique shops.[53] Nikolaus Pevsner and Edward Hubbard were insistent that Chester is a Victorian city. 'In the popular view Chester is the English mediaeval city par excellence [...] Chester is not a mediaeval, it is a Victorian city. What deceives is the black and white. 95 per cent is Victorian and after.'[54]

This passage, nevertheless, illustrated one significant root of Rolt's approach to historic environments, artefacts and practices. Rather than pursue enumeration, measurement and assessment through documentation, he relied on impressions and intuition, sometimes acute, sometimes (as it would turn out) inaccurate. He expressed a profound sense of prospective loss, perceiving continuities with historical worlds whose survivals could be glimpsed but were fragile and fleeting. This would extend to concern about the impact of industrialisation on the loss of crafts, which he viewed as more in tune with the natural world. His interest in craft would develop into a strong practical bent, which would lead him to empathise more with history that was oriented around practical observation and deduction than with theoretical work that leant heavily on written documentation and ideas.

The route that the family walked on Christmas Eve (newspaper weather records suggest that it was 1916) can still be followed. While additions have been made to the cathedral since that time, the main structure remains that 'restored' from the medieval (with two preceding attempts) by George Gilbert Scott between 1868 and 1876. Although Rolt would comment adversely on many Victorian restorations (which inspired the founding of the Society for the Protection of Ancient Buildings), he did not cite the Chester restoration. Even in 1868, this had been attacked as a rebuilding rather than a restoration.

The family party must have followed Northgate Street to the Cross, then a crossroads with two main roads that linked England to Wales on either side of the Dee, but with little or no motor traffic. The street has since been partly pedestrianised, with Watergate Street paved with 'heritage' setts in place of cobbles. It remains possible to climb the same steps to Watergate Row, which was historically more neglected than other rows until conservation works began in the 1960s. The first section includes various doorways that gave access to small workshops behind the front facades, most of which were in poor order and subsequently swept away. At least one of these led to a closed Chapel Court, fated for demolition in 1957, when other courts, further down, had already been demolished.[55]

Halfway down is the mouth of the 'narrow alley, dark as midnight', which leads through to the back of Watergate Row. If this alley is followed to an access road, which opened up the rear area in the mid-1960s, it is clear that, behind the facades, the demolition of parts, serving courts, workshops and slum housing has removed the sense of enclosure.

Beyond the passage, much has been rebuilt, including a whole section to replace one that collapsed before demolition in the 1950s. This, 'the Gap', was reconstructed in concrete after 1968. While the form of Watergate Row has been partly preserved, much of what the young Rolt saw has disappeared, although impressions remain.

Watergate Street itself meets the dual ring road, created by demolition and the widening of Nicholas Street in 1964–66. A long Georgian terrace to the west side of the street has survived. This was known as Pill Box Row, after the many medical practitioners there at the start of the twentieth century, although it seems that Dr Taylor's surgery was on the opposite side of the street and thus demolished. As the ring road follows the line of the former Roman wall, what lies beyond lay outside the Roman city.

Crossing Nicholas Street, the first road on the right is Grey Friars and at the end of this, on the left, is the former Greyfriars House, at which Rolt stayed on childhood holidays. 'For an impressionable small boy, there could have been few more romantic houses, and certainly no more romantic city, in which to spend Christmas,' he wrote later.[56] While this had been encased in brick in the Georgian era and heavily altered, it was a single large residence with servants quarters when Rolt visited. After being saved from demolition in the 1970s, it is now apartments.

This marked the end of the Christmas Eve walk, but it is worth following the Walls of Chester opposite the canal. The road here flanks the city walls above the racecourse, with views of the Dee and railway viaduct beyond, where Rolt used to watch trains. Following the Walls north over the Watergate, the site of the former infirmary for Chester – at which Dr Taylor officiated – is passed. While its historic central section has been retained, this now comprises residential apartments. The Walls then cross a long bridge over what was formerly the London & North Western Railway. It may be surprising that, although Rolt later recalled taking trams to Chester General Railway Station, he made no mention of this bridge or indeed the canal scene beyond. One possibility is that this was seen as too recent: the widening of the railway (piercing an even wider section of the Walls) and the rebuilding of canal bridges and canal-side walls is dated to 1902–05, less than two decades before he explored the Chester area.

Ahead, reached down steps, Tower Wharf can be seen on the Shropshire Union Canal and in the distance is the former Shropshire Union yard at which narrow boats like *Cressy* were built. *Cressy* itself would be built at Pontcysyllte the year after Rolt's Christmas Eve walk.

After the Walls cross the railway, the wide basin at Tower Wharf can be seen to the north, and steps lead down from the Walls to the canal towpath near the foot of Northgate Locks, following this to pass beneath Raymond Street. If the young Rolt explored this, he did not mention it later, but this may well have been seen as a disreputable and dangerous place for an upper-class boy. He certainly never travelled here by boat, but he did call at the office here in 1947 to obtain a permit to go up the canal towards Llangollen.

On the far side of the wide basin are recently erected student flats, built on the site of what had been a field for boat horses. On the towpath side is a historic dry dock, still in use, and a slender 'turnover' bridge that allowed horses to cross the canal to take up the towpath on the far side. It is from this bridge, looking back towards the Walls, that most of the visiting boats, the locomotive and Alvis car could be seen at the 2010 celebration.

Chester was the only city that Rolt knew well. In a sense, the city that he depicted (and later, perhaps, Nantwich) represented the acceptable urban form – of seemingly ancient origins, rooted in the rural and of historic character.

The historic house of Greyfriars in Chester, where Rolt stayed during holidays as a child.

The turnover bridge at Tower Wharf, Chester, where the towpath passed from one side of the Shropshire Union Canal to the other. The fence on the left guards the dry dock here, while the large plaque commemorating Rolt's involvement is on the side of the bridge.

Detail of the plaque to Tom Rolt on the turnover bridge at Tower Wharf, Chester.

Watergate Street in 2016. The Rows, on which the Rolt family walked 100 years earlier, are on the left. Note the varied styles and ages of buildings: Watergate Row has been much rebuilt.

'Occasionally the mouth of a narrow alley, dark as midnight and leading who knows where, opened up.' It seems that this was the main passage leading through from Watergate Row, but others gave access to courts.

# 4

# THE EARLY INLAND WATERWAYS ASSOCIATION AND WATERWAYS REVIVAL

Between 1945 and 1951, Tom Rolt was heavily involved in the founding and running of the Inland Waterways Association (IWA). In his writings, he had advocated and endorsed the idea of major social transformation, without proposing many practical measures that could help achieve this. To use writing to seek to change the climate of public opinion, as Massingham's numerous works seemed to propose, may have constituted one incidental aim of *High Horse Riderless*. However, it was the publication of *Narrow Boat* that would lead to a movement for change to revive British inland waterways, collectively and individually. Rolt's extensive participation would lead him towards the two years in which he was the manager of the Talyllyn Railway, which involved multiple practical activities alongside much paperwork.

The elegiac tone of *Narrow Boat* provided no indication that it could prompt a national campaign for the conservation of waterways. While it lamented the possibility that much would disappear, and the book itself perhaps provided some form of 'conservation by record', it put forward no practical means to help conserve an intangible heritage. The idea for a pressure group (of sorts) did not come from Rolt, but from his co-founders, who saw some potential if public opinion could be mobilised. This chapter will discuss some aspects of his involvement, including his expulsion by Robert Aickman from the organisation that he had co-founded and to which he had devoted five years of unpaid labour.

The definitive account of the IWA, employing all available sources, has yet to be written – either of the first twenty years or more recent times.[1] However, there are several detailed accounts of the first five years, often focusing on the differences between the personalities of Rolt and his co-founder, Robert Aickman[2]. While these interpersonal problems have some significance, the history of this pressure group – its successes and failures – owes much to the environment within which it operated and the way in which it developed among possible alternative models.

## The Founders of the IWA

*Narrow Boat* had perceived the impending losses of a delicate world, but not how these might be halted. Among the many readers who wrote to Rolt, two would suggest that something could be done.

Charles Hadfield (1909–96) favoured the development of modern waterways for freight transport and tourism. Then seeking nomination as a Labour MP, he was unimpressed by many of Rolt's doubts about the modern world that was emerging from wartime. His experience of Aickman led him to withdraw from active involvement soon after the IWA was formed.

Robert Aickman (1914–81), who would become the dominant figure in the ensuing campaign for waterways revival (and remained so for twenty years), was the other. The orphaned, only child of an architect, during wartime, Aickman had become involved in a literary agency, Bloomsbury, seemingly one in which his wife, Ray, did most of the work.[3] He occupied a peripheral position in the middle class, like Rolt's but with different perspectives. He shared some of Rolt's concerns about the 'collectivist bureaucracy' that seemed to threaten people like them – especially in the world of elite art, theatres and opera, which formed an important part of his relatively privileged life. Although Rolt and Angela were introduced to parts of Aickman's world, Rolt does not seem to have been greatly influenced by his ideas. He continued to follow many of the views that he had developed with Massingham, but perhaps with growing despair.[4]

Both Rolt and Aickman shared a disdain for the world of the 1940s, especially the post-war reforming Labour government; they saw much of what they valued as in decay and close to extinction. Against this, Aickman thought 'it possible that a fortress could be built, which could not easily be either stormed or assimilated', through a 'social adventure' that would involve the revival of waterways.[5] This, it seems, could initiate the more widespread social change that would restore many aspects of the hierarchical elite society into which he had been born.

Rolt would strongly emphasise attempts to retain an intangible heritage: support for a dying work culture associated with carrying by narrow boat. This meant traditional canals, working methods, boats and, above all, people.

Contrary to much that has been written, they would part company over Aickman's domineering intransigence rather than any clash of outlook or policy. There were significant differences, but at first, after the two met on *Cressy* at Tardebigge in August 1945, theirs was a reasonable meeting of minds.

The British middle class, in 1945 as now, was divided in various ways, and that included the extent to which it seemed threatened by the rise of Labour and social democracy. Many in the middle class realised that much of their social world would survive, albeit without domestic servants and with a more unionised working class that might gain some privileges that had previously been confined to the middle class. For some, the pre-war period was seen as one from which everything had declined. Aickman felt that the worlds of privileged wealth – deferential service to unequivocal superiors, holidays in high-quality hotels, leisurely but secure well-paid work (his father had lived in a private club before his marriage) and an elite that enjoyed music, literature and theatre – were all under threat.

Rolt's background also lay in relative wealth, but his regrets were not over the loss of a secure, metropolitan, middle-class way of life. He focused on the threatened loss of craft and the rural and traditional industries. Both he and Aickman shared the same distrust of machines and mechanical civilisation, so that Aickman would commend *High Horse Riderless*. These are, perhaps, odd perspectives for the two leading people who helped to found a popular environmental pressure group, especially in the 1940s. Hadfield, who soon withdrew, was much closer to the mainstream.

## Pressure Groups and Purposes

Pressure groups, which were less numerous in the 1940s than today, could follow varied models and methods. Some would have or aspire to gain the ear of government, seeking to move policies towards directions more acceptable to their supporters. These policies might be based on expert research and reports or simply promote and stress established positions.

Other groups would take an oppositional position, aiming not so much to influence the direction of government policy as to substantially oppose and change it. This could be through research, publications or lobbying, or they might pursue protest in a more-or-less-organised fashion. In many ways, the IWA that emerged would be closer to the latter. Under Aickman, it would not advocate different policies for government-owned waterways but sought to replace government ownership with what he termed a National Waterways Conservancy.[6]

The context within which the new pressure group would operate was the nationalisation of many inland waterways – including almost the whole network explored by Rolt for *Narrow Boat* – under a public corporation handling much inland freight transport. This did not abolish the existing legal framework whereby waterways had to be maintained to minimum standards and with a public right of navigation (until the latter was removed by the Transport Act 1968). Nationalisation took effect from 1948, but Rolt recognised that it would be necessary to deal with the new state owners if the IWA was not to be stillborn.

The form of public ownership and the duties of the state-owned body could have varied from those that the new British Transport Commission (BTC) would adopt. Nationalisation was able to offer some prospects of public investment in waterways development, although it would become clear that this would only apply to a small number of larger waterways. Little emphasis was laid on this during the early days of the IWA, although publicity was given to J.F. Pownall's bizarre scheme for a 'Grand Contour Canal' – an inland ship canal.[7] In principle, it would have involved the revival of some waterways for leisure use, which predominates today but was then extremely limited, and it could have provided significant support for the beleaguered narrow-boat carrying industry and especially the small numbers who lived on board 'family' boats. The latter had drawn Rolt's admiration, but while there would be some support, the overall approach of the BTC and its subsidiaries (beginning with the Docks and Inland Waterways Executive (DIWE)) was to press for modernisation and the elimination of what were seen as archaic methods, vessels, practices and routes.

The new association would have to draw up policies based on limited knowledge and research. At its very earliest stage, Hadfield advised that research be carried out into the current position and, on publicity, 'The more we know, the more press publicity we can get'.[8] The IWA would need to pursue varied but limited cooperation with what were always called the 'authorities'. It would have perhaps expected to find allies, develop relationships with managers of waterways and associated facilities, and to expand membership and enthusiasm among the general public.

To achieve this, it would need an efficient organisation, the involvement and use of enthusiasts on the ground, internal and external communications (press and publicity), and campaigns that yielded positive results. If the IWA of the 1940s did not have success in these respects, this was partly due to the manner in which it was founded and developed, rather than any failings of specific individuals.

## Founding the IWA

Much depended on the kind of group that was envisaged while the Pacific War was still under way; whether this was the right moment must be debatable, but the UK government was already considering policy for the post-war world. In his first letter on 9 July 1945, Aickman asserted, 'It has long occurred to me that some body might be founded to promote the welfare of the canals and general interest in them', and he suggested an interest/pressure group.[9] This would occupy an uncertain position between the kind of supportive 'friends' organisation of supporters and more vociferous groups campaigning for political change. Examples of the latter would emerge, partly in opposition to the post-war Labour government. For instance, one body very much on the political Right was the British Housewives League, a protest group formed in 1945 to oppose austerity, the welfare state and food rationing.[10] However, the aims proposed by the group after Rolt and Aickman met at Tardebigge proved modest enough.

At the outset, Rolt and Aickman planned to act as promoters, aiming to found an organisation that others would administer. Rolt's list of fan letters about *Narrow Boat*, and Aickman and Hadfield's limited contacts would be used to get this started. From the outset, it was envisaged that paid staff, led by a salaried organiser, would do the work, leaving the founders as figureheads. This was a familiar enough model for many voluntary organisations of the time and indeed later. The initial aims were somewhat unclear: they mainly sought to preserve and develop waterways throughout Britain, although the early focus rested on the Midlands and southern England.[11]

Aickman's letter cited three organisations that any new society could follow. One was the Friends of Canterbury Cathedral (FCC), which had been founded in 1927 by the dean and chapter of the cathedral and was described as a 'body of supporters'.[12] Before nationalisation, there would be multiple waterways owners and authorities with whom to deal; in contrast, the FCC sought to assist a single owner 'in the care of the cathedral and its fabric today, and in its preservation for posterity'.[13] This was distant from Aickman's later desire for a private organisation to take control of the waterways and run these in an unspecified general interest. Similarly, he mentioned the Glyndebourne Opera, although this was a private operation that had begun at its operator's home in 1934.[14]

The other body cited was the Light Railway Transport League (LRTL), which was perhaps closer to what the IWA became – an often vociferous pressure group. Aickman was a member between at least 1946 and 1951.[15] The LRTL had been founded in 1937 in response to proposals to withdraw trams from the north London area. It sought to protect existing tramways and light railways and promote new ones. 'Speeds up to 100 miles per hour' were suggested for new lines.[16] The LRTL protested about the replacement of trams by trolleybuses without much success, but it claimed to have the support of 'tram managers, workers in the tram industry, the manufacturers' and the general public.[17]

Under the Aickman-dominated IWA, few in the waterways industry – owners, carriers or workers – would openly support this pressure group. The LRTL settled down to the promotion of public exhortation, visits to tram systems and, increasingly, farewell tours. London's trams would all be withdrawn by the summer of 1952.

A subsidiary activity of the LRTL was to study the history and operation of historic tramways and preserve historic trams. This led to the formation of a breakaway organisation – the Tramway & Light Railway Society – and eventually to the Tramway Museum Society in 1955. The IWA presented no real equivalent, although other organisations had begun to consider the preservation of historic craft.[18]

Rolt himself would develop approaches to waterways history, while Hadfield embarked on studies in history that would lead to a comprehensive series in the 1960s.[19] However, none of this involved formal organisation, and even the founding of the Railway & Canal Historical

Society in 1954, which Rolt would join, provided canal history only as an afterthought.[20]

Rolt had a different model for the founding: the VSCC. However, this was really a members' club that included relatively wealthy enthusiasts who had their own vehicles, camaraderie and access to the Prescott Hill Climb in Gloucestershire. As he would recall, this club, founded in 1934, had never engaged in internal controversy (or indeed external problems). From 1945, he was also involved in the Newcomen Society, upon which the IWA's initial constitution would draw. These were bodies of enthusiasts, with a strong scholarly element to the Newcomen (founded in 1920) that studied historic industry and transport.

## Methods and Scope

Early correspondence indicates much doubt over methods. Rolt was clearly the driving force, with Aickman talking but making no practical progress.[21] Hadfield considered that the original aims, as drafted by Rolt, were modest: 'It tends to give the impression that the battle is already lost – for instance, your remark that "the promoters are opposed to the whole conception of central and collective planning". Personally, I am in favour of it.'[22] Hadfield laid much emphasis on publicity, encouraging pleasure boating and walking on towpaths. Rolt agreed with most of this, and pressed Aickman to convene an early launch meeting.[23]

Perhaps influenced by his interest in the River Severn, Rolt envisioned environmental concerns that were wider than navigation on canals. He suggested the title 'Inland Waterways Society'.[24] Having discovered that there was a Pure Rivers Society in London, he suggested collaboration, or better still, 'some form of amalgamation'.[25] Both he and Aickman considered that affiliated bodies such as the National Trust and CPRE might have representation in the new association.[26] Other organisations, such as the Ramblers Association, did join, but mostly for the mutual exchange of literature – there was no direct representation.

The association was duly launched in May 1946 with a small but growing membership that needed to be serviced with publications, meetings and events. Many members were actual or aspiring pleasure-boat owners, some forming their own circles – later formalised into branches – that focused on their local waterways. Aickman was the self-appointed chairman, who supplied his premises for meetings and a headquarters office, while Rolt became the secretary to whom Aickman sent most of the correspondence for action. While this arrangement was meant to be temporary, it would continue for nearly four years of Rolt's life.

This provided a framework for the development of a pressure group, but one that rested on shaky foundations. It relied greatly on the personal efforts of the two leaders, while Aickman's autocratic personality made it difficult to delegate any work or accept any sort of internal democracy. It also put considerable pressure on both men; such dependency has often proved to be a feature and weakness of many voluntary organisations.[27]

## Administration and Problems

A great deal of correspondence, meetings and member-oriented events ensued. However, it would not be until 1949 that serious paid assistance was obtained, and with an organiser who was soon dismissed.

Neither man intended this, and Rolt would set out his misgivings in a series of letters to Aickman, notably on 8 June 1948. While he had seen life on *Cressy* as a means of escape from commitments, and enabling him to write, he was increasingly constrained by the need to report his movements on a regular basis and to spend much time on administration. While Aickman was able to escape into a personally congenial world of sorts, Rolt was being drawn away from his own world: 'the pleasure of life on "Cressy" — the lack of distracting concerns and the absence of the feeling of being bounded by time.' Rolt was a private figure in many ways but was being forced to attend outings and events that, unlike Aickman, he did not enjoy. As early as February 1947, Rolt had expressed concerns — he was then halfway through the writing of *Green & Silver*, his book about inland waterways in Ireland, and would have four books hanging fire.[28]

Few would suggest that a full-time honorary secretary of a pressure group, handling large amounts of correspondence (most of which was forwarded from Aickman's home), should do so from a travelling home on a boat, needing to make intermittent calls at post offices to collect *poste restante* mail. Similar problems applied to one of the few commercial boaters who became involved, Sonia and George Smith, who were involved in carrying on the Grand Union — messages had to be delivered through diverse means.

Work for the IWA was also causing financial difficulties for Rolt. He had foregone his Civil Service income after 1945, intending to make a living by writing. In the spring of 1947, he told Aickman, 'I am becoming increasingly concerned about the state of my finances', while the expense involved in Aickman's proposed six-week cruise of IWA officers through the waterways of northern England was jeopardising a promised short return to Ireland. 'I must confess that in my straitened circumstances [...] the sum involved is causing me the deepest concern.'[29]

The administrative work only continued to expand and, by 1949, Rolt was complaining that he was working on IWA matters and writing for seventeen hours a day, seven days a week.[30] On 3 December 1949, he finally told Aickman that he intended to retire from the secretary position.[31] Aickman was, in his own words, 'aghast', but while Rolt expressed regret that this announcement 'should have caused such consternation and despondency', he stressed that he had always intended to drop out once the IWA was firmly established. While Aickman liked being a 'public feature', 'it is a position which I find most uncongenial'.[32] This led to a long period in which relations deteriorated, but the seeds had been sown long before. Out of support for waterways and their associated life, Rolt had been assigned a role that he had never sought. Had the IWA developed so that there was sufficient income to employ staff (as it would later), his role would have been very different.

Rolt contended that the organisational tasks and requirements failed to fulfil his needs. 'What I need to counterbalance my literary activities is some practical and manual activity,' he would write later. He meant something like the design and joinery work that he had carried out to convert *Cressy* to a home and maintain it.[33]

## Meetings, Campaigns and Achievements

What were the achievements of the organisation in the five years since it was founded, until Rolt and others were expelled?

It is instructive to consider a later submission, which was made to the official Bowes Committee inquiry into waterways in October 1956, over five years after Rolt had been expelled.[34] The IWA leadership explained its campaigns and apparent achievements, but it is notable how limited the latter had been. The submission focused on four waterways for pleasure or commercial use: the Kennet & Avon, the Stratford Canal, the Essex & Suffolk Stour and the Derby Canal. While there had been publicity as these navigations continued to be unnavigable, only the first two would be restored throughout: the Stratford in 1964, the Kennet & Avon in 1990. In the latter case, legal actions had been launched, but these resulted in compensation for affected traders, not restoration.

The 1956 submission provided no list of successful campaigns and omitted failures, like the Basingstoke purchase and Huddersfield Canal voyage, both in 1948 and against which Rolt had mildly counselled. In other cases, opposition to the Rochdale Canal closure (1952), the Barnsley Canal (1953) and Stroudwater Canal (1954) were cited, but these had all been unsuccessful. The replacement of the nationalised authority by a 'National Waterways Conservancy', a major idea of Aickman, was advocated.

What was not provided, beyond a reiteration of the Lifford demonstrations (q.v.), was a list of solid practical achievements. The Lower Avon revival, which was proceeding with much voluntary labour and financial input, was not mentioned. Until 1949, when Douglas Barwell, a new member of the IWA Midlands Branch, took over, Rolt had taken responsibility for the campaign to revive the Lower Avon.[35] Whether the IWA would have been more successful had Aickman not sought to divide it is unclear. It is notable that many of those who would be expelled were keen supporters of the Kennet & Avon revival.

For any voluntary campaigning organisation, much depends on the environment of policy and practicalities, and this was difficult in the extreme. Many but not all of the inland waterways were nationalised in 1948, so that any engagement seeking publicity and support from the former private owners would have involved dissipated efforts. It is not clear whether much contact was made with the railway companies, upon which much blame was attributed for the decline. The Great Western Railway was one exception; it had sought to close the Kennet & Avon Canal just before nationalisation and had allowed navigation over the northern Stratford Canal, upon which traffic had ended, to be impeded by the fixing of an opening bridge crossing at Lifford. After the well-known Lifford Lane (actually Tunnel Lane, Lifford) protest cruise on *Cressy* in 1947, Aickman was concerned to find that Parliament and the new BTC regarded this as merely an internal operational matter.[36]

The lack of knowledge about and membership in northern England led Aickman to propose a long voyage through waterways in the area. There is little evidence that this caused much positive publicity, but it had quite damaging consequences for the Huddersfield Canal. Despite closure in 1944, this remained just about navigable, being maintained to ensure water supplies. After the hired boat was sunk in the second lock of this canal in autumn 1948 and the canal was drained, water supplies to a local mill were disrupted, with ensuing layoffs. The new nationalised owners then removed lock gates to make further passages impossible.

One waterway that was too lightly trafficked for nationalisation was the Basingstoke Canal in Hampshire and Surrey. When this came up for sale by auction, attempts by some IWA members to secure acquisition were outflanked by a party that claimed IWA support but instead set up a new company that did not prevent it from falling further into dereliction.

More general policies would depend on pressure – by convincing argument or opposition – upon the new administrations formed by nationalisation. A meeting in March 1947 with Minister of Transport Alfred Barnes was followed by one in March 1948 with Sir Cyril Hurcomb, Chairman of the BTC, although Hurcomb proved to have little interest in a small section of a very large new undertaking.[37] At both meetings, the memorandum that Rolt had prepared was pressed. This had sought an independent inquiry, with a new policy to cover all canals while the abandonment of any waterway should only follow careful investigations.

At the meeting with Hurcomb, Rolt had made a special plea for the future of the canal to Llangollen, which he had navigated in 1947 in *Cressy*. He felt that this navigation, which had already been abandoned in 1944, should be retained for pleasure use and the plans for local authorities to culvert road crossings resisted. Later, Aickman and Rolt met Sir Reginald Hill, who chaired the DIWE, which managed the BTC waterways (and docks).[38]

Whether any of these figures would have been impressed, whatever the policies urged by the IWA, is doubtful – they had their own masters. The model the BTC followed was that of the Morrisonian public corporation, owned by government but treated like a private body that made its own decisions. While it began with the principles of integration – for instance, road transport collecting and delivering from ports, waterheads or railheads – in practice, these separately administered transport bodies competed with one another.

At the same time, much trade had been disrupted. In wartime, some traffic had ended but, despite new temporary movements, little new traffic was developing on the smaller waterways. Some of the largest carriers were being absorbed into the nationalised organisation – for instance, the

Grand Union Canal Carrying Company (and other subsidiaries). When Fellows Morton & Clayton went into liquidation, much of its operation, boats and employees was absorbed into the BTC south-eastern fleet. Smaller carriers, notably the much-vaunted 'Number Ones', were being squeezed out.

Aickman and Rolt soon found that support for waterways was equivocal: the trade body, the National Association of Inland Waterways Carriers, proved to be more concerned with retaining good relations with the new state owners than campaigning with a small voluntary organisation for the revival of less-used lines.[39] While S.E. Barlow, owner of one of the two Barlow coal-carrying companies, would form one exception, little support came from canal owners or carriers.[40] This bears out the comments in Frank Pick's wartime report, which found that most carriers were inward-looking. This had to be true of the carriers whom Rolt most admired – the independent owner-boatpeople – many of whom could not read and write, and whose ability to support or negotiate for new traffic development was necessarily limited.

One illustration lies in the failure of the Severn Carrying Company (SCC) to publicise the Worcester & Birmingham Canal, over which it still carried by narrow boat. Asked in 1948 why its publicity brochure did not include the canal, SCC general manager E.W. Bayliss responded, 'We feel, however, that it is quite unnecessary for us to give any explanation as a private Carrier to your Association for our actions'.[41]

## Boaters and Publicity

Before the IWA was formed, Rolt had been involved in the Ealing semi-documentary, *Painted Boats* (1945). It is probable that the filmmakers had noticed an article by him in *The Field* in 1944. The article had been very pessimistic, noting major closures that were then proposed, including many of the Shropshire Union lines. Rolt, blaming railway company ownership, asserted that 'many of the canals have already become stagnant ditches, and more are likely to follow them'.[42] Acting as a consultant, he helped to ensure greater authenticity, although his practical commentary was replaced by a more 'arty' one by the poet Louis MacNiece.[43] Although this was a fictional account of two boat families, it used real boaters and raised interest among many viewers, as well as providing a visual record.

The IWA had some success in gaining support from private leisure-boat owners and hire-boat owners (although when the IWA was founded there was only one canal-based hire firm – the Inland Cruising Association, near Chester). *Narrow Boat* had spawned reminiscences from those who had used pleasure boats before 1940 and interest by others in acquiring and using such boats.

The IWA had an early plan to acquire its own boat for use by association members. This did not succeed, and Aickman found little use for his own boat, which he had acquired with Rolt's assistance. The driving practical force here was Rolt, who was able to appraise possibilities and comment on the sometimes extortionate prices being sought by the vendors of private boats. However, many existing and new boat owners did not prove to be especially active: many wished simply to own and use their boats, but leave it to others to press for the retention of waterways for boating.[44]

It was Rolt who prepared advice for boaters and provided lists of boatyards and, as they developed, hirers (the 'pink list', which was sold to members).[45] This helped to service members but also to support and develop a network of waterways for pleasure. He advised new hire firms, notably Rendel Wyatt from Liverpool, who formed the Canal Cruising Company at Stone on the Trent & Mersey Canal in 1948; the IWA *Bulletin* had usefully carried details of two boats that Wyatt acquired.[46] Later, Rolt befriended Holt Abbott of Saul (and later Stourport), who designed and built a prototype pleasure boat, *Avondale*, which could go anywhere on English canals; he accompanied the Rolts on their way to the Market Harborough event in 1950.[47]

It is notable that in some ways, this went against Rolt's emphasis on the retention and revival of waterways for carrying, but he was one of very few who had enough knowledge and practical experience of boating. An early document produced by him, 'Commercial and Pleasure Traffic on Canals', focused on problems that pleasure boating could cause for carrying boats. Among many DON'Ts, he exhorted boaters: 'DON'T moor at wharves or other points where it is obvious from the well used rings or other evidence that working boats are accustomed to lie. Loaded boats on account of their deep draught can only moor conveniently at certain points whereas you have a wider choice.'[48]

There were some successes with publicity. In 1947, the Rolts set up a public exhibition, which was staged first at Heal's furniture store in London, facilitated by Tom's friendship with fellow VSCC member

Anthony Heal; later, it was toured to various locations, publicising waterways.[49] Using his experience with pre-war car rallies for the VSCC, Rolt suggested a national 'Rally of Boats', to be held at Market Harborough, where pleasure boats from the Trent could easily reach (including, perhaps, those from the Derby Motor Boat Club of which he had so disapproved in *Narrow Boat*). Scheduled for the summer of 1950, this idea was unilaterally extended by Aickman to host a full-blown literary festival. Rolt was dismayed by the enlarged requirement for voluntary effort that this would necessitate, but the rally would raise much public interest.

The limitations of commercial trip boats — there was a fleet at Llangollen and a short-lived operation on the Lancaster Canal — led the IWA to run its own trip boats. Rolt took a leading part in ones in London, despite his doubts about cities; these provided a means to develop membership but also to publicise waterways in urban areas.

From the outset there had been plans to issue a monthly magazine, which Rolt would have edited, to publicise the cause.[50] This never appeared; at the AGM in April 1948, Rolt explained the many problems involved. These included paper shortages, the doubtful profitability of a magazine and the need to rely upon voluntary involvement when existing helpers were already greatly stretched.[51] The *Bulletin* featured limited production values (duplicated from stencils), lacked illustrations and was increasingly dominated by Aickman, who literally dictated most of its contents from his Bloomsbury office. Until branches began to be formed, this became the only communication for members. Its tone became shriller and more vituperative, perhaps as the realisation grew that the IWA's successes would be limited in an increasingly unfavourable environment.

It must be stressed that others made significant contributions, but it was Rolt who had most of the ideas and attempted to put these into practice when possible.

Six years after the publication of *Narrow Boat*, Rolt had spent five of those dominated by voluntary involvement, something that would be followed by two years with the Talyllyn Railway. He would later complain that being identified as a writer meant that authors could be deemed willing to assume all sorts of honorary roles. This reflected some of the perils of involvement in voluntary causes.

## Specific Campaigns and Narrow Boats

Within the limitations of his travels, Rolt took early responsibility for three campaigns about individual waterways. The Derby Canal had been partly disused since the late 1930s, with a final traffic in 1945. Its private owners had sought to close it and redevelop parts of its site. They had offered it to the new nationalised authority in 1949 without success. In 1945, one director of the Derby Canal Company, B.A. Mallender, had voted for closure but then led opposition to it from late in 1946. After *Cressy* was refused access in 1948, ICI, a former customer located east of Derby, was persuaded, with Rolt's support, to order a load of coal to be brought by canal from the Cromford Canal via the Erewash Canal. The loaded narrow boats duly arrived at the Derby Canal junction but found the entrance locks padlocked. The coal was then offloaded to lorries and delivered at the Derby Canal Company's expense.[52] The view that there was a right of navigation that could be enforced was here thwarted by the payment of compensation, something that would also take place with legal proceedings over the long-term failure of the Kennet & Avon Canal to be properly maintained.

In the case of the Welsh Canal (now known as the Llangollen Canal), Rolt had travelled in *Cressy* up this canal – which had been closed in 1944 – in both 1947 and 1949, and found local opposition to plans to substitute low road crossings for bridges. Little could be done bar exhortation and publicity, but here he found support from Christopher Marsh, the BTC Divisional Waterways Officer for the north-west, who successfully presented an argument that navigation would need to be maintained to ensure the passage of maintenance vessels. Rolt paid tribute to this in *Landscape with Canals*.[53]

Narrow-boat carrying was clearly in decline in the 1940s, and once the possibility of intervention was raised, Rolt sought to support it without fostering too many intrusive changes. The unresolved issue of living-in on family boats – which many had opposed – presented one issue. Rolt identified that a major reason why some were leaving narrow-boat carrying was the need to ensure a reasonable education for their children – about 250 children lived on family boats.

Some existing waterways owners fostered decline. Rolt felt that the Grand Union Company, in particular, was carrying out anti-competitive practices, while the Severn Carrying Company had run down carrying on the narrow Worcester & Birmingham, employing poor-quality, inexperienced crews. Better trade practices, handling methods, education (minimal at least), rationing and pastoral support and better relations with waterways owners and representation, through unions if necessary, were all possibilities.

A further activity would be to encourage and support carrying, but it is difficult to see how a small group could achieve this. One possibility might have been to seek out opportunities for new traffic (or the prolongation of existing ones), but the main approach became to observe and deplore the loss of traffic.[54]

It was Rolt, along with boaters Sonia and George Smith, who felt that subtle enhancements to the lot of existing boat crews, along with improvements to the canal track, handling and practices, might help to conserve this way of life. Other possibilities included better administration, such as a licensing scheme in place of tolls. Few such workers joined the IWA, but a Working Boaters Sub-Committee was formed. This included meetings with officials from the Transport & General Workers Union. Although Rolt was averse to unions on principle, he was surprised to find himself impressed.[55] Attempts were made to improve rationing for workers and press for improved dredging to enable larger and more reliably navigated cargoes to be carried.

Much of this had to be behind the scenes, as with, for instance, Rolt's attempts to get the support of the S.E. Barlow coal-carrying firm. Nationalisation of some carrying might have presented a possible path to support, but it soon became clear that a selective rundown was envisaged as a necessity. However, by the time that the Working Boaters Sub-Committee had reported, the expulsion of several members was under way.

## IWA Splits and Expulsions

What had the IWA achieved by the end of 1949, when the divide between Rolt, Aickman and their supporters began?

There was publicity for what had been an obscure cause and, ironically, given his aversion to personal publicity, much of this was down to Rolt. The Heal's exhibition tour and forthcoming national boat rally provoked much interest. There was, however, limited influence over major actors in the waterways – the BTC, its subsidiary DIWE, major carriers and other canal owners proved to be unimpressed.

Solid achievements on the ground/water lay outside the central IWA leadership. Supporters of Linton Lock in Yorkshire, in the IWA North-Eastern Branch, began to raise funds late in 1949 to help restore this lock.[56] The Lower Avon revival and restoration, originally proposed by interests in Evesham, was furthered greatly by Douglas Barwell, then a new member of the Midlands Branch, and his personal acquisition of that waterway in 1950. Only the Tunnel Lane protest proved successful after several boat movements forced the obstruction to be raised.

In the 2020s, a common account seems to have arisen that the 'saving' of the waterways was due to 'volunteers'. This seems to involve an apparent confusion between those who do voluntary work with those who form and run pressure groups like the IWA and later canal societies. A recent academic book refers to 'the volunteers who saved the canal network from ruin and who continue to care for it'.[57] By 1951, the only practical work by volunteers on a British waterway was on the Lower Avon, on which working parties had recently begun.[58] Volunteers of course joined and took part in the IWA, and its leadership were volunteers, but the impression that voluntary work was beginning to save the waterways is difficult to sustain.

What might a general pressure group hope to achieve in its first four years? Some measures would include increased membership to bring together those who might be interested, developing greater membership numbers and member involvement. This was partly achieved by the formation of branches, beginning with the Midlands Branch, North-Eastern Branch and Kennet & Avon Branch. Successful campaigning would bring publicity and the encouragement of pleasure boating and perhaps for carrying by boat. The latter, however, depended less on public support and more on cold financial appraisals of the economic possibilities of the retention or expansion of carrying.

The IWA would be weakened by Aickman's attempts to remove members who dissented from his views. He became increasingly dictatorial and negative after Rolt announced his departure, such that Rolt walked out of an IWA Council meeting on 17 June 1950 and formally resigned on 14 July. His resignation letter pointed to disagreements with Aickman's insistence that the IWA was 'a wide cultural organisation' and his tendency to make decisions without IWA Council approval.[59]

In an exchange of letters, Rolt stated that he had often acted as a buffer for the resentment that Aickman could arouse, and he had now 'abandoned the most invidious position it has ever been my misfortune to occupy'.[60] There was unpleasantness over the Market Harborough

Festival, and Aickman then made it clear that he sought 'a clearer definition of our aims', in order to secure explicitly 'so far as possible the elimination of those who disagree with them'.[61] He noted that dissenters such as Lord Lucan and the carrier John Knill (a Kennet & Avon campaigner) were more interested in supporting carrying, whereas Aickman claimed to seek to support all waterways in every aspect.

This formed a wedge that Aickman could drive in so that the IWA would, quite deliberately, be split. He promptly organised a Special General Meeting (SGM) for 30 December 1950 in London to affirm a resolution against any policy of 'priorities' and to agree new rules. This meeting was to break up in confusion.

This prompted a small group of dissenters who circulated the IWA membership with their concerns. Rolt recorded that Rendel Wyatt, the leader of this group, called on him in Banbury early in 1951, 'like some gunpowder plot conspirator', and he added his signature to a memorandum that was critical of the changed constitution and Aickman's leadership.[62] The signatories were varied: Wyatt had started a hire-boat firm, while others represented support for carrying. They do not seem to have represented a coherent alternative that could have formed a solid cohort for a rival society.

Aickman's resolutions were passed at a reconvened SGM in Birmingham on 3 February 1951, amid further confusion. By publicly threatening to resign if these resolutions did not pass, he made this a vote of no confidence. Subsequently, much effort was devoted to the expulsion of the dissenters through dubious legal means. The other two co-founders, Rolt and Hadfield, were expelled from membership, along with others, notably those associated with the Kennet & Avon campaign and with carrying by enthusiasts.

Was this necessary, and did it further the aims of this pressure group? There is little evidence that any damaging divide was developing so that a purported general IWA cause was under threat. Members no doubt joined to support various aspects of the cause or local waterways without disavowing any general cause. Those who raised funds for Linton Lock in Yorkshire were local boaters who supported the IWA in general, but their local waterways cruising grounds most; they might be less moved by threats to a Kennet & Avon navigation that their own boats could not reach.

Aickman insisted that any language of 'priorities' would suggest that support would be focused on some waterways at the expense of others, and 'the waterways stand and fall together'. This was only one interpretation: an alternative was to focus attention and resources on some waterways and activities first, and others later, but not to abandon the latter.

This had been going on in previous years. Rolt had pointed out as early as December 1947 that perhaps too much attention was being paid to the Kennet & Avon.[63] Meanwhile, carrying on the Grand Union was declining, with the family boats under threat. In February 1949, Rolt had suggested that there was little future for the Rochdale Canal: 'There are certain waterways over which we should not shout too loudly lest we acquire the reputation of seeking to keep any and every canal in being.' Aickman had not reacted then with fury, but responded mildly that no waterways should be abandoned.[64] It is difficult to discern whether there was anything to gain from this declaration, except to increase Aickman's power over the organisation and the removal of any dissent. It certainly served as a means of revenge against anyone who challenged his views.

Rolt was already trying to sell *Cressy*, but its hull was found to be severely rotted beyond repair and it was abandoned. He would soon transfer attention and commitment to the Talyllyn Railway, and received the expulsion letter from the IWA while manager of the railway.

The reduced IWA would go on to be moderately successful in generating public support for a waterways revival. There were clearly flaws from the start: an uncongenial policy environment; lack of support from parties in the waterways industry or local or central government; a chaotic position with unpaid volunteers, including Rolt, running a growing national organisation; and, despite much publicity, a failure to secure effective campaigns. The oft-quoted Lifford protest was one of the few successes, with the bridge replaced and traffic returning to part of the Stratford Canal in 1952. Other successes depended upon voluntary money (the Linton Lock and Lower Avon purchase) and the beginning of voluntary labour (the Lower Avon) in locally led campaigns.

Ironically, the IWA succeeded in encouraging the growth of pleasure boating; although Rolt was doubtful about this, his leaflets helped to expand interest, vessels and facilities. There was less success in his major concern to protect working boat people, especially the family boats. There were few representatives among working boat people themselves, and the advocacy by the Working Boaters Sub-Committee was stymied when Aickman expelled leading members. Over the latter, it is hard to see how far effective campaigning and policies could have saved the family narrow boats – Rolt himself later admitted that, at most, this could have delayed their end by ten years.[65]

What if Tom Rolt had declined to be involved in the formation of the IWA or had left in its early stages? It may well be that this organisation would have soon foundered, if it had been formed at all. His writing skills enabled him to be an effective honorary secretary, despite the disadvantages of living on a mobile boat; he could even type his own letters, whereas Aickman required a secretary. Tom's specialised knowledge of waterways was unmatched among enthusiasts in this period. This included much practical knowledge, which, for instance, enabled him to advise prospective boat hirers and boat builders. It would be difficult to identify anyone who, even if they had his levels of enthusiasm and commitment, could have filled this role. In that sense, not only did he inspire many who read *Narrow Boat*, he followed this up with great commitment and unpaid efforts while he sought to make a living through writing. To be expelled was scant reward for so much valuable effort.

This and *Narrow Boat*, not the limited achievements of the early IWA, perhaps represented his most significant contributions to the future of Britain's inland waterways. It must remain a matter of speculation as to what would have happened to those waterways had the IWA failed to be formed, or lasted a very short time, which might have happened without Rolt's efforts. Much may have been lost or opposition to closures might have arisen in a different manner later – and quite possibly too late.

Despite his mild aversion to cities, Tom Rolt helped to run day trips for IWA members, to gain publicity and encourage membership. This photo shows George Smith (boatman and Sonia Smith's husband at the time) steering *Cairo* into Islington Tunnel on the Regent's Canal in London on 23 April 1949. Rolt is standing on Smith's left. (Ian L. Wright, courtesy of Stephen Rowson)

# 5

# TO THE RURAL IDYLL? IRISH WATERWAYS AND RAILWAYS

In 1946, while the IWA had just been formed, the Rolts spent the summer in the south of Ireland. Tom was partly intrigued by the (possibly apocryphal) story in Smiles' *Lives of the Engineers* about the 'shoemaker' who defied the Grand Canal Company directors and promoted an alternative line from Dublin to the Shannon. There was no obvious candidate, bar a former shoemaker involved in both the Grand and Royal Canals – the major ones in the south of Ireland.[1]

Angela suggested an exploration of these waterways in 1946 and a possible visit to the Scottish canals (presumably the Forth & Clyde) in 1947, and she would later suggest that they should acquire a larger boat to explore waterways in mainland Europe.[2] Tom felt gloomy about England in the immediate post-war period and had thought of 'retreat to Western Ireland. I can't help feeling that Ireland may prove to be the last citadel of humanity but I may be wrong.'[3]

Ireland seemed to embody a form of rural idyll, and he would find this in some ways, some of which were unanticipated. He wrote later of rural Ireland, 'I was irresistibly reminded of the simpler, more natural world of my childhood on the Welsh border'.[4]

Before Rolt became fascinated by inland waterways, he had been moved by railways, especially narrow-gauge ones.[5] This may well have begun in infancy, with a glimpse of the locomotive *Katie* on the Duke of Westminster's private Eaton Hall estate railway, and reinforced by visits to the Glyn Valley Tramway and the ailing Welsh Highland Railway.[6] He would, of course, go on to co-found the movement to revive the narrow-gauge Talyllyn Railway in Wales. *Green & Silver* relates some railway travels alongside a long journey over waterways in the south of Ireland in the summer of 1946. Unlike the voyage recounted in *Narrow Boat*, this did not relate to a period in his life residing on his own boat, but a lengthy holiday with a clear end. While *Cressy* was due for major repairs at Banbury, Rolt chartered a hired converted lifeboat, *Le Coq*, from Athlone for three months in 1946. This enabled the Rolts to explore parts of the south of Ireland at leisure.

While his account is that of an outsider, it would inspire interest by boat owners (mostly on the Shannon) and waterways enthusiasts in Ireland, some of whom would later found and support the IWAI. The Ireland that he found had been affected by shortages and restrictions in the Emergency years – as the early 1940s were known in the neutral Irish Free State – including the rationed supply of fuel for the boat that he had hired for three months. Ironically, Rolt praised the use of horse-drawn transport in Ireland, although this had only become necessary due to private motor-fuel rationing since 1942.

## Influences and Inspirations

Rolt's involvement with various influences was as a reader and writer; he did not set out to expound others' visions or views, but in considering and visiting Ireland, sought out and stressed aspects of this thinking. Visiting only twenty-five years after the civil war in Ireland and just as the Emergency had ended, he might have written more closely about these features of Irish history. However, what he emphasised was influenced by organicism, his readings of William Butler Yeats and probably Taoiseach Eamon de Valera, and his interest in historic transport. This did not prevent him from commenting about aspects of Ireland of which he disapproved, but that was not his central purpose.

While one inspiration behind the new book was the poet and politician Yeats, it is probable that Rolt was aware of and maybe even heard the address by de Valera on *Raidió Teilifís Éireann* on St Patrick's Day, 17 March 1943. He may well have felt some resonance with de Valera's references to Ireland as:

> The home of a people who valued material wealth only as a basis for right living, of a people who, satisfied with frugal comfort, devoted their leisure to the things of the spirit – a land whose countryside

would be bright with cosy homesteads, whose fields and villages would be joyous with the sounds of industry, with the romping of sturdy children, the contest of athletic youths and the laughter of happy maidens.[7]

It seems doubtful that while Rolt often visited and highlighted places that were in manifest decline, he would not favour policies to encourage a revived agriculture, industry linked to the rural and farm workers, and 'peasants' engaged in leisure 'of the spirit'. His descriptions of Richmond Harbour at the northern end of the Royal Canal, where he found most buildings in a ruinous state yet with the Cloondara Mill still working nearby, would provide one example.

De Valera's address has been derided as his 'comely maidens' broadcast, although, whatever its meaning, he never used this phrase in the broadcast. It has also been seen as idealising a romanticised Ireland that did not exist; similar accusations could, as earlier chapters have indicated, have been made against Rolt. However, de Valera did not set out a vision of Ireland as it was, but as it might be ('The Ireland That We Would Have') against a modern, urban-industrial society, with landscapes, language and beliefs to match. The broadcast did not reflect a manifesto for the future development of the Irish economy and society. It marked the 50th anniversary of the founding of the Gaelic League, which primarily sought the revival of the Irish language.

The Gaelic League had been founded by the Protestant Douglas Hyde to preserve and promote the Irish language, but it soon become nationalist and republican in practice. Like others, de Valera saw the restoration of Gaelic as evoking an earlier, de-Anglicised Ireland, which could be achieved in the future and could form a lodestar of sorts for advocates of republicanism. Similar visions seem to have inspired British politics in the 1939 war – fighting for an England that was profoundly rural, pastoral and under threat, despite the fact that most of the population, unlike Ireland, lived in urban settlements.

Some evidence for war and the English pastoral was provided by Peter Scott, later an IWA vice president, in an Easter Day broadcast in 1943:

> I suppose the 'England' means something slightly different to each of us [...] probably for most of us it brings a picture of certain kind of countryside, the English countryside. If you spend much time at sea, that particular combination of fields and hedges and woods that is so

essentially England seems to have a new meaning. I remember feeling most especially strongly about it in the late Summer of 1940 when I was serving in a destroyer [...] I thought of the Devon countryside [...] and I thought of the mallards and teal which were rearing their ducklings in the reed beds of Slapton Leigh. That was the countryside we were so passionately determined to protect from the invader.

As Wright has commented, 'Did Peter Scott really go to war for a duck? – as the nation is reinvented around the imperatives of the present situation'.[8]

It would be surprising if Rolt demurred at de Valera's vision, although the Ireland and state of 1946 was very different from this rural utopia. The main tenets around which policies clustered were summarised, just before they were jettisoned in 1957–58, as 'Protection, self-sufficiency, keeping people on the land, retaining industry in Irish hands only; these sacred Fianna Fail tenets all went out the window'.[9] As the poet Seamus Deane contended, 'By 1932 [...] the alliance between an anti-modernist Church and an introverted state and culture had been consolidated. It was, thereafter, to reach paranoiac proportions at times.'[10] In context, these had underlain economic depression and massive emigration, most notably from the rural west.

Of the Ireland that Rolt encountered, the Irish historian R.F. Foster asserted:

Much in 1930s rural Ireland would have been recognizable to a reincarnated Victorian traveller. Housing remained dominated by the single-storey cottage; living conditions were basic; families large; emigration and tuberculosis part of life [...] There were some improvements in housing, but in 1946, only 5 per cent of farm dwellings had an indoor lavatory.

Farming mechanisation was represented only by tractors. Foster attacked de Valera: 'His ideal, like the popular literary versions, was built on the basis of a fundamentally dignified and ancient peasant way of life.'[11]

This raises questions about Rolt's perspectives prior to and during his long visit. He had clearly expected an impoverished country in which services were inefficient and government control significant, although not to the level of control in wartime Britain, but with a more secure survival of a traditional rural society approaching some sort of idyll.

## Planning *Green & Silver*

This provides some of the background to Rolt's visit. It should be emphasised that this was not directly inspired by his readings of Yeats, and probably reports of the de Valera broadcast. But they may well have been in his mind, providing images that he found favourable, fostering expectations that might be met or confounded. It was not a literary visit, but one upon which he would reflect. He had already written *Narrow Boat* and *High Horse Riderless*, and the former had, of course, been published. He probably had publication in mind for *Green & Silver*, which forms a careful account of the journey. In style it is far less Arcadian than *Narrow Boat* and more like a factual travelogue that now provides a useful historical document, albeit one in need of careful interpretation.

His journey began in June 1946, arriving by steamer to Waterford.[12] Having expected the mainline railways by which he reached Athlone to be antiquated, he was pleasantly surprised: there is some sense of order in his commendation of neatly painted locomotives and country railway stations, contrasting with the grime of those in Britain. He portrayed an efficient system rather than one that had survived from the past, as is suggested in his later accounts of the narrow-gauge lines. The first is a report by a tourist using an effective service; the second is the fascinated enthusiast finding an historical attraction. It was somewhat ironic that it should be a railway employee, Jack Beahan of Athlone, who owned *Le Coq*, on which he and Angela would spend the following three months.

He perceived a sense of reduced pace, summed up by 'time enough'. He dismissed the somewhat Weberian (Protestant ethic) notion that this might have an ethnic basis in 'the shiftlessness of the Irish race abetted by a mild, relaxing climate', against 'the bustling and efficient Anglo-Saxon'. Contrasting the hurrying of crowds in English cities, 'I wondered whether perhaps we have lost the measure of progress and of greatness' and 'who then shall say that he is not more truly civilised?'[13]

## Exploring Railways

Rolt had experienced mainline railways to reach Athlone but would incorporate a narrow-gauge long-distance line into the final stage of his journey through Ireland. This was the West Clare line, involving a 53-mile journey. He took a first-class ticket on the only train that day, in 'a compartment that was a period piece in itself', commending its fittings. Traversing the wild, agriculturally poor country through which it passed, he noted that local people ('the primitive, weather-beaten people of West Clare') crowded station platforms not to travel, but just to see the trains.[14]

They changed to a single carriage train at Moyasta Junction, and so on to Kilrush. He wrote that he would have been sorry to miss this trip, but feared that before long, 'unless the public taste for travel changes, it may no longer be possible to repeat it'. The railway had just been nationalised but its possible replacement by road services had been explored in 1945, while conversion to 5ft gauge was rejected; the line would last until 1961.

He followed a more typical branch line – one of several narrow-gauge lines owned by Córas Iompair Éireann (CIÉ) and used mostly for freight only – with a journey along the Cavan & Leitrim Light Railway, which had only one passenger train daily. He noted that the locomotive was original, built by 'Robert Stephenson of Darlington', as were the two passenger coaches. They passed numerous level crossings on the way and then changed to the section that formed the Arigna Branch, which was more like a tramway running alongside the public road.

It must have been clear that these lines had very little future, and this may have influenced his feelings about the Talyllyn Railway in Wales, which he had already visited. Without use by tourists, these could not survive.

## Carriers, Carrying and Coverage

*Narrow Boat* was centred around carrying by narrow boat, but all waterways in Ireland were navigated by wide boats. The book had expressed much sentiment over the family narrow boat, and especially over the owner-boat people 'Number Ones'. Rolt began by commending the Grand Canal Company, which would shortly be nationalised and placed under the control of the railway-dominated CIÉ. This would be similar to the process in Britain, with waterways under a railway-dominated BTC, but at the time of his visit this was not yet in place (or, indeed, its implications known).

The Grand Canal Company, which also ran boats on the Shannon and Barrow, claimed to 'operate a greater mileage of inland waterway than any other company in the British Isles'.[15] This was probably true, yet the nearest English equivalent was the Grand Union Company, which owned and operated over the London–Birmingham and Derbyshire routes. Both companies operated road haulage – the Irish company on a larger scale. However, the chairman of the Grand Union in the 1940s, John Miller, had advocated its modernisation and the development of road fleets to and from 'roadheads'. Ironically, Rolt (and others) would come to condemn the trade practices of the Grand Union, whose carrying subsidiary tended to compete away much trade from the 'Number Ones' that he so admired.

He found some resonances with the family boats in that, although each boat was crewed by three or four males, the most junior member of the crew was often the son of the captain. The son carried out both domestic tasks and the setting of locks, but 'one day he will become a captain like his father, and so the skill of the boatman passes to successive generations'.[16] This would not prove to be the case, as all traffic would cease in 1960. He noted the dangers posed by the lack of proper landing places at locks and the deaths that resulted.

Rolt provided brief details of the two remaining carriers on the Royal Canal. There are inaccuracies in that he met Patrick Caffrey (not 'Cafferty') in Dublin and, later, James Leech (not 'Leach') of Killucan. His encounter with Caffrey portrays him as an eccentric; he seems to have sought to have acted as guide to the visitors but they never saw Caffrey again. They met Leech's boat, carrying bog ore to Dublin, and frightened the two horses that were drawing it – unlike the motorboats on the Grand, the canal banks on the Royal could not have withstood the erosion created by powered craft.

As with *Narrow Boat*, informal conversations with boat people proved informative – in this case, a former Royal Canal boatman whom Rolt met at Jamestown on the Shannon. Once again, there seems to have been no systematic collection of historical details.

The Royal was pictured as a canal in little use and in much decline, albeit kept in reasonably navigable order by the railway company that had long owned it, and whose weed-cutting gangs kept most parts clear. The section west of Keenagh proved to be out of use, with 'increasing signs of decrepitude', and when they came to the lower lock at Killashee, 87 miles from Dublin, it seemed doubtful that they would be able get through.

One balance beam on a top gate had broken, while the paddle racks were missing. Water was leaking through the lower gates so that the lock could not be filled, but they eventually managed to lever up the paddles on the broken gate, 'whereupon the level in the lock chamber began slowly, desperately, to rise. We tore up clods of turf from the lock sides and threw them into the water above the bottom gates where, sucked down by the current, they staunched some of the leaks.'[17]

Watching the water levels in the lock chamber rise very gradually, they finally forced open the sound gate, while *Le Coq* was narrow enough to be able to enter the lock without the need to open the second gate.

As with *Narrow Boat*, there were comments about religious buildings and places en route. During a prolonged stay at Athlone, they visited the recently completed Church of St Peter and St Paul. Rolt was unimpressed by the exterior: 'The great neo-Classical pile with its white walls, columns and campanile, and its dome of black Galway seems to have no roots in the Irish landscape, but to have been translated bodily from some sunnier southern climate.' However, he found the interior 'dignified and impressive. The reason for this is simple: the whole exhibits a sense of design and of craftsmanship. It is not just a nondescript building filled with the stock-in-trade of the Dublin church furnishers. Ireland could do with more craftsmen, particularly in her churches.'[18] One of the churches in Tullamore was filled with 'Catholic repository bric-a-brac', while the Protestant one was locked and the 'interior, damp, fusty and decaying', with only swarms of bees in attendance.[19]

On their first day on the Shannon, south of Athlone, they anchored *Le Coq* at the ancient site of Clonmacnoise, where Rolt was horrified to have to beat their way through nettles to find older gravestones smashed to make way for 'a sea of unsightly tombstones of marble or polished granite' and a wooden hut with a 'jerry-built altar'.[20] However, they found some contemplation of the wider life on this ancient site, in a manner similar to his response to Llanthony Priory:

> No landscape can have changed so little in a thousand years. In these days of chaos, arrogance and confused thinking it is a pity, I thought, that more men cannot contemplate in quietness such innumerable solitudes [...] They make man aware of his creaturehood, of the brevity of a life 'bounded by a sleep', and of the vanity of ambition. But while it thus humbles him, the natural world enlarges man's humanity by enabling him to perceive the potential greatness of the human spirit with its unique creative capacity.[21]

When they reached Maynooth on the Royal Canal, they moored and visited the village and college there. Now part of the National University of Ireland, the latter was then a very different institution, training cadres of Roman Catholic priests. In comments that would later draw ire, Rolt asserted, 'At his best, the parish priest is a patriarch in the truest and finest sense; at his worst he is an extortionate petty-dictator ruling by superstition and fear'.[22]

He went on to acknowledge his limited knowledge of Catholicism but felt that he should 'attempt to define my own reactions to it'. He contrasted the dominance of the Church with that of bureaucracy – a familiar theme in his writings. He seemed to conflate the growth of bureaucratic rule in Europe with the threat of Communism. This triumvirate seemed to be part of 'the contagion of our world disease', and 'both Communist and priest display a puritanical and masochistic attitude to life which results in the loss of the humanities and the eclipse of the arts of living'.[23] He concluded that, as a Christian, albeit an Anglican, he would reluctantly have to come down on the side of the priest in a conflict between Roman Catholic and Communist, but he accused the Church of stagnation and dogma. These pronouncements were, perhaps, compounded by his view of the architecture of the college and criticism of the early Victorian architect A.W. Pugin, an exponent of the Gothic Revival style. Rolt preferred the earlier work at Maynooth.

*Green & Silver* paid greater attention to the natural world than *Narrow Boat*. Issues like the cutting of fuel from the turf bogs were discussed, alongside their encounters with wild and semi-wild swans on the Royal Canal and a grey squirrel near Lough Boderg. Of the latter, he commented that this 'destructive, rat-like beast' had now invaded Ireland from England and 'will ultimately exterminate that most endearing of all small animals, the red squirrel'.[24] He wrote with awareness rather than a demonstration of expert knowledge.

## Leisure and Boating

*Green & Silver* places much more emphasis on leisure use than *Narrow Boat*, despite the complete lack of pleasure boats observed on the Grand and Royal Canals: both were seen by their owners as a means of transit from Dublin to the Shannon. In turn, limited numbers of boats were noted in transit between the Loughs on the Shannon, on which much of the leisure use was local only. But this was a time of petrol rationing, in the first summer when rations might be available.

Perhaps reflecting thoughts about the beginnings of the IWA, the possibilities of pleasure boating are discussed, especially the provision of secure harbours in towns. He recorded problems with local children on the Grand Canal and made a very early start in order to ascend the Royal Canal out of Dublin. The attention of local children proved a problem on the Royal Canal when they moored at Maynooth, and theft affected the boat when it was left in apparent safety a mile west, at Jackson's Lock.

The Shannon provided the focus for pleasure boating, with much emphasis upon the major navigational dangers over the large loughs en route. This included confusion over the exact navigable route, fog and the effects of wind and flooding. Navigational travails with the boat and dinghy are recorded – something that is absent from *Narrow Boat*.

Rolt's encounters with Lough Derg Yacht Club were positive. He and Angela were welcomed immediately and were invited to participate in sailing races. Many invitations to individual lakeside homes followed, and it was to his regret that they could not stay longer in Ireland to enjoy further welcomes.

His hosts were mostly from Ascendancy backgrounds, being Protestant and Anglo-Irish and, like the English boat people, they met with approval:

> Farming, riding, fishing, shooting, or sailing, despite all the troubles with which Ireland has been rent, these people continue to follow a way of life which can have changed very little in two hundred years, but which has virtually disappeared from England as a result of our bloodless but far-reaching social revolution. In a remote and sparsely populated countryside, hospitality has not been strained by abuse as is the case in our overcrowded island. To cross their thresholds was for me a bitter-sweet experience, unlocking forgotten stores of the nostalgic recollections of a childhood when vestiges of such a life still lingered in England and then seemed to me stable and assured.[25]

The last sentence is a critical passage, expressing much behind Tom's outlook – a nostalgia for an Edwardian England that he had glimpsed.

Clearly, the Rolts presented an acceptable face to pleasure boating, although it was mostly not the kind of holiday touring that would offer a future to the English, and indeed Irish, connected system of waterways.

Rolt had come across another leisure user further north on the Boyle River, a tributary of the upper Shannon. Having moored *Le Coq* on Lough Key, it was swept away by wind and rescued with the assistance of Sir Cecil Stafford-Harman of Rockingham House, which overlooked the lake. Invited to dinner at the house, Rolt described Rockingham as 'an urban mansion set in the wilds of Connaught; it has an arrogance which is incapable of any concession to its surroundings'.[26] He found Stafford-Harman congenial, quoting William Cobbett's approval of the 'resident native gentry, attached to the soil'. Writing of 'the evil of absentee landlordism', Rolt asserted, 'The Irish peasant lacks the local guidance, encouragement and example that a good landlord could provide'.[27] Quoting Yeats' lament for Coole Park, he concluded, 'If the landowners of Ireland had looked upon the health of their land and its people as a personal responsibility and not as a source of exploitation, such a beneficent hierarchy might well have developed'.[28] His sympathies clearly lay with the neo-medieval order of benevolent landlords and grateful peasants.

Rolt had seen press mention of the formation in 1945 of the Shannon Development Association (SDA) by Monsignor Timothy Joyce of Portumna. A Grand Canal agent in Portumna arranged for them to meet when he reached the town. Joyce favoured the restoration of the pre-1918 steamer service on the Shannon and the improvement of its amenities, but he had also pressed for more buses in the Portumna area and the development of tourism, centred round the river.[29] Rolt praised his 'vision of the present as a synthesis of past and future'. Sadly, the SDA does not seem to have long survived Joyce's death in February 1947.[30]

## Visiting Dublin

In Dublin, Rolt managed to persuade a reluctant Grand Canal Company to permit *Le Coq* to be moored in Ringsend Docks after it refused consent for him to moor at Portobello Harbour, on the Circular line. Although it had allowed him to moor at its enclosed harbour in Tullamore, the Grand Canal Company did not permit mooring in Dublin, and he commented on the need for safe moorings in cities for pleasure boating. He advocated 'the establishment of centrally situated commodious yacht basins adequately protected from inquisitive intruders and equipped with those facilities such as drinking water which the water traveller needs'.[31]

While moored at Ringsend, the Rolts and their guests, the Hodgkinsons, 'did most of the things that visitors are expected to do'.[32] His comments on their visit to the Abbey Theatre are worthy of note. He was unimpressed by the buildings ('a gloomy and depressing theatre') and by the play they saw, which was almost certainly *Professor Tim*, a 1925 play by George Shiels; this play was one that was often produced by amateur companies. 'The bucolic comedy of Irish rural life which we saw there, though it was amusing enough and competently performed, was hardly of the standard which one expects of a national theatre which the Abbey now claims to be,' Rolt observed. This bears out Declan Kiberd's later comments:

> Dublin was overrun by unplanned migrations of rural folk, who had no sooner settled than they were consumed by a fake nostalgia for a pastoral Ireland they had 'lost' [...] while the Abbey Theatre which, located in the heart of Dublin's north inner city, continued all too often to play the rural classics for tourists and (increasingly) for the natives. The Abbey for long periods functioned more as an artistic museum than as an experimental or a people's theatre.[33]

Yeats, one of the founders of the theatre in 1903, had written in his posthumous *On the Boiler*, 'The success of the Abbey Theatre has grown out of a single conviction of its founders', it was what the dramatists wanted, not the public. Yet the theatre has not, apart from this one quality, gone my way or in any way I wanted it to go.'

Rolt rarely commented on city environments but did so in *Green & Silver*. He condemned the newly sprawling growth of suburban Dublin, 'like our own cities', at the expense of rural areas:

> Affluent Dubliners escape to fashionable suburbs while the beautiful Georgian houses of the Squares become offices or degenerate into tenements [...] Today the bulk of our population exist in these sterile suburbs where the arts of living die, where a wedge is thrust between that once intimate relationship of town and country, and where the one is looked upon as a workshop and the other as a playground.[34]

He commended the broad streets and squares brought about by the Wide Streets Commissioners from 1757 'and the genius of the eighteenth-century planners', but condemned modern planning. He asserted that the creation of Georgian Dublin happened 'under the inspired patronage of men who were the heirs to generations of good taste and sensibility, and who knew the arts of living in cities'.[35] These upper-class wealthy men knew how to employ architects and builders, but there was no sense in which there could be modern equivalents from a 'semi-illiterate bureaucracy' of planners.

As with the older rural England, with industry at craft level, he advocated the same for Ireland, despite the lack of craft skills. His admiration for craft in England was not matched by that in Ireland and he commented that he must conclude after his tour that he had seen 'little talent for craftsmanship of any high order. While I had seen much solid honest workmanship, I had also seen much rather shiftless bodging.'[36]

However, in Ballinderry he had found Dick Stanley, a maker of violins. He was a local baker who had learned through intuition rather than any training.[37] Rolt did not draw any conclusions about the revival of craft and its significance in providing a base for a sustainable rural economy in Ireland.

In *Green & Silver*, he began to suggest how the tide of change might be turned back or deflected, endorsing the small scale advocated by de Valera and organisations like Muintir na Tíre.[38] That better material, if not spiritual, lives would result from improved economic conditions (despite some losses in amenity) was dismissed.

## Reception and Conclusions

How was the book received after its eventual publication in 1949? An otherwise positive review in the *Catholic Standard* (30 December 1949) condemned 'some blundering generalisations about Jesuits, Maynooth, and Irish parish priests'.[39] The same review, by one who knew 'every stretch of the water of his journey from Athlone up and down the Shannon', had anticipated familiar errors by Englishmen writing about Ireland. However, *Green & Silver* proved an exception, and 'here was the near-perfect travel book about our country'. The reviewer agreed with Rolt's comments about Clonmacnoise and the modern Church of St Peter and St Paul in Athlone.

As with *Narrow Boat*, one consequence was publicity for Irish waterways and the eventual formation of the IWAI in 1954. This had built on concerns about the Shannon, the proposed fixing of low-opening bridges and the restoration of the steamer service proposed by Monsignor Joyce. It may well be that the general accuracy aroused interest that would not have applied to a less well-informed account by an outsider.

Rolt would return to Ireland, partly because 'in rural Ireland I was irresistibly reminded of the simpler, more natural world of my childhood on the Welsh border' — a theme to which he would often return.[40] He revisited in 1952 as part of the writing of *Lines of Character*, but this time to record the Tralee & Dingle Railway. Later, in a new edition of *Green & Silver*, he commended the increase in pleasure cruising, but 'no canal or river can ever be quite the same when their commercial life is dead, when the low-laden craft are seen no more and once busy wharves grow silent'.[41] This could, perhaps, also form a conclusion about most waterways in Britain — the track and engineering conserved, but not the working environment for which they had been constructed or the corresponding lives.

Heavily gated off in this photograph from 2017, Tullamore Harbour is where the Hodgkinsons joined the Rolts to pass through the eastern Grand Canal and the Royal Canal.

Sherriff Street Lift Bridge on the Royal Canal in Dublin looked 'like a pair of slender, long-beaked birds'. It had only recently been installed when *Le Coq* passed through to moor beyond, just north of Spencer Dock. This is where they met Patrick Caffrey, who offered to pilot the boat through for them.

Maynooth Harbour, where *Le Coq* was moored while the Rolts went into what was then the village of Maynooth. The railway station in the background featured a signal box, where the signalman kept children at bay from the boat. The station closed to passengers the year after the Rolts' trip but was reopened in 1981. The slipway, part of the interim restoration of the Royal Canal, does not look like it is used very often.

*Opposite:* On the Grand Canal, Rolt managed to moor at the canal-side village of Robertstown. This view was taken in 2007, when the former hotel (in the distance) had not long closed after several years of community use. When Rolt moored here it was used as a Turf Board hostel. An auction in March 2024 proved unsuccessful. Note the Barge Inn on the right.

# INTERLUDE
# TARDEBIGGE

To visit Tardebigge, where *Cressy* was moored between 1941 and 1946, is to explore two strands of Rolt's involvement with waterways: the founding of the IWA and his growing feelings about waterways and their place in landscape. The second is reflected in his memory that 'these Tardebigge years turned out to be one of the most happy, full and fruitful periods of my life [...] for me a time of rapid spiritual and intellectual development'.[42]

Rolt's own description in *Landscape with Canals* of the Tardebigge that he encountered can hardly be bettered. He saw it as one in which 'architect and civil engineer had collaborated with nature to provide for Cressy a landscape setting which suited my taste to perfection'.[43] To this was added the 'little community of craftsmen' at the maintenance depot. As the canal was under threat at the time he was writing about it, with the Pick Report recommending closure, Rolt's account could have formed an elegy. Much would have been lost.

It may well be best to begin a visit next to St Bartholomew's Church and school, with its convenient car park, but also its views from the churchyard over the canal and over as far as Bromsgrove. The distant church spire that Rolt mentioned is still visible, albeit surrounded by suburbs that have grown since the 1940s. The dramatic setting of the canal, emerging from the short Tardebigge Tunnel onto an embankment before it begins its descent by thirty locks, can be clearly seen.

A clear, well-surfaced footpath, including a fenced section, drops down from the car park to the canal towpath, with the tunnel mouth visible and with clear views of the depot opposite. Rolt viewed the latter when it was in decline with, for instance, the final boatbuilder in place. By contrast, in earlier times there had been up to twenty boatbuilders on the canal, with tunnel tugs serving the tunnel at Tardebigge and three other tunnels on the canal's summit level. In *Worcestershire*, Rolt sadly recorded the death of Tommy Hodges, the last boatbuilder; there was no replacement.

Walking towards the lock, past the depot buildings and moored pleasure boats, it is worth turning to view the hill through which the tunnel was bored. This had a single main road, running to Bromsgrove, but this has been replaced by the A448 Bromsgrove highway, a dual carriageway which has, to its credit, been landscaped with limited visibility into the hill; however, the traffic noise is noticeable and makes this a much less quiet place than in Rolt's time. Another view, from the footpath leading back to the church, provides glimpses of both roads on two levels.

A substantial lock-keeper's house stands next to the deep Top Lock; after a period in which it was commercially let, it was recently sold. Rolt knew a lock-keeper, Jack Warner, from further down the flight – a regular at the Halfway House, a small canal-side pub that closed in 1963. In turn, this property was sold in 2018.[44]

Crossing the lock with care, the point at which *Cressy* was moored until 1946, on the offside embankment, is marked by two plaques. Ironically, mooring is no longer permitted here, and on my latest visit, late in 2023, two Canal & River Trust maintenance craft were moored here. Interestingly, in wartime there was much suspicion locally about Rolt, who set off by road each morning to his employment with the Ministry of Supply.[45]

The plaques record the place at which the two main founders of the IWA, Robert Aickman and Rolt, first met. The Aickmans had walked from Bromsgrove Railway Station, although now there is a bus service. The first plaque, from 1981, records their meeting as 1946; the second, unveiled in 2005 by Rolt's widow, Sonia, corrects this to 1945. Had this meeting not taken place – if Aickman had not read *Narrow Boat* – it is doubtful that this canal, at least, would have survived. Commercial interests in the Birmingham area had abated losses since the 1920s and it is probable that this subsidy would have ceased.

The path can be followed past a series of lime kilns, comparatively recently excavated, and the boatyard and drydock; the latter, from 1924, has been occupied by boatbuilders Crafted Boats Ltd since the 2000s. This is now historic in its way, having been founded in 1953 as J.L. Pinder & Sons, converting narrow boats – the successors to *Cressy*.

The use of the depot has been much reduced since Rolt moored there, and it has been rumoured that it might close altogether. Rolt did not mention that several of the buildings here dated back only to 1909–11, when maintenance was transferred further north from Stoke Prior, joining the tunnel tugs that had been stationed here since 1874. George Bate's carpenter's shop dated from 1909; Tom Insull's smithy from 1911.[46] These were recent buildings when Rolt visited them in the 1940s, yet the craft work that they housed was much older, perhaps dating back to 1811, when this part of the canal opened.

Walking past the workshops, one of the historic tunnel tugs, *Birmingham*, from 1912, is on display on the bank; another, *Worcester*, is preserved at the National Waterways Museum at Ellesmere Port. Around the depot was a small residential community, including the superintendent's house and a post office, and over the main road, the Plymouth Arms. The early closure of the latter – on temperance grounds (it is now a nursing home) – and the loss of the working community and subsequent sales, makes this more a residential enclave than one oriented around the canal. If this had been the position in the 1940s, Rolt would not have derived so much inspiration from it and would have had to seek it elsewhere.

The church footpath can be accessed by walking over the tunnel portal and back to the towpath. Much appears similar to the Tardebigge that Rolt knew, but its functions have changed. If the depot was to close and its buildings were redeveloped, the people and place that inspired him would inspire no longer. At least the whole area has escaped the fate for which it might have been destined in the 1940s: closure, draining and destruction of the depot buildings. This was the fate that its former owners, the Sharpness New Docks Company, favoured and it nearly took place under nationalisation.

The general view from Tardebigge churchyard, with the lock cottage and *Cressy*'s wartime mooring in the centre.

A general view of Tardebigge depot buildings in 2015, with the tunnel cutting in the distance and the depot buildings and warehouse on the offside.

Tardebigge Top Lock looking towards the depot and tunnel cutting in the distance, with *Cressy*'s wartime mooring to the left.

The brick pillar and stone bank mark the site of *Cressy*'s wartime mooring in 2015.

Another view of the brick pillar and stone bank on the offside, which mark the site of *Cressy*'s wartime mooring in 2014. Tardebigge Top Lock and lock cottage are in the distance.

Detail of the plaque placed in 2005, which corrects the date when the Aickmans and Rolts first met to 1945. Unveiled by Rolt's widow Sonia, it records that it was, in fact, Robert Aickman who supplied the inaccurate date!

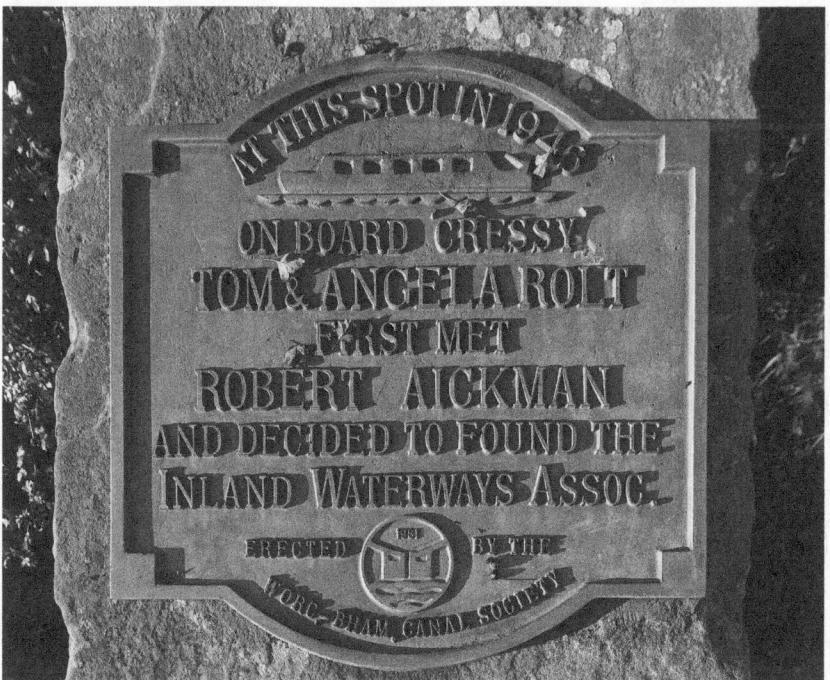

Detail of the plaque placed in 1981, which inaccurately recorded 1946 as the first meeting of the Aickmans and Rolts, and omits Robert Aickman's wife, Ray.

Canal & River Trust maintenance craft moored at the site of *Cressy*'s wartime mooring in 2023.

Approaching Tardebigge Tunnel in 2015.

Tardebigge workshops, with the former warehouse, now in residential use, in the distance.

Tardebigge tug *Worcester*, completed in 1912 and preserved at the National Waterways Museum at Ellesmere Port following restoration by the Boat Museum Society.

# 6

# WORCESTERSHIRE, CRAFT AND WATERWAYS

Rolt's third book was not specifically about transport. He was commissioned, in unusual circumstances, to write one as part of a series of county books edited by Brian Vesey-Fitzgerald. Vesey-Fitzgerald had been editor of *The Field*, for which Rolt had written. As he was then living on *Cressy* at Tardebigge, he chose the volume on Worcestershire, although this was not a county that he knew particularly well outside its waterways. He seems to have been heavily influenced by his own book, *High Horse Riderless*, continuing to make some apocalyptic assertions. In this case, Rolt related his own ideas to something tangible: the history and topography of a county.

The landscapes created by partial industrialisation and urbanisation and rural survival were matched by the intangible heritage of people who contributed to those landscapes. He focused particularly on surviving craftsmen.[1] The book strongly reflects his view of craftspeople as critical driving forces mediating humans and nature, consciously evoking the idea of ecology at a time and to a readership when this term was rarely used outside technical discourse.[2] As in *High Horse Riderless*, he felt that civilisation had taken a wrong turn, and elevated the surviving craftsmen to a model that should be followed. The tenuous and delicate nature of that survival, and its possible consequences, was stressed.

This linked to his idea of history which, in this case, relied on the 1939 book by Robert C. Gaut, *A History of Worcestershire Agriculture and Rural Evolution*. He viewed this as a mine of information 'of a sort never to be found in the orthodox County History'.[3] Curiously, volumes of the *Victoria County History* for Worcestershire had been produced; the fourth volume, which included Pershore, had been published in 1924. Gaut was described as Worcestershire County Council Agricultural Organiser from 1914 before his retirement and death in 1941.[4] His background was not in history, however, but in the study of agriculture.

Gaut's study underlay Part One of Rolt's study, 'Time', on which he freely acknowledged its dependency.[5] This approached a form of environmental history: 'The standpoint I have chosen has been the ecological one. I have tried to demonstrate that the Worcestershire we know and admire is not the work of nature alone or of man alone, but the product of a fruitful partnership between the two.'[6]

This accounts for the idiosyncratic form and style of his account, in a county series that proved very uneven and was soon superseded. The text continued to embody many apocalyptic assertions, such as:

> Because this deep-felt desire and love which each man feels for his own place has been exploited, twisted and perverted by this unhappy age into a militant nationalism, our intellectual socialists would destroy it, and, in their ignorance, pour mankind into their arbitrary collective mould. Such an abstract concept cannot succeed, and it were better that the world of men should founder in chaos rather than it should do so, for in that case man would become a homeless migrant and, more than that, he would lose his soul.[7]

His conclusions expressed the fear of the 'gigantic machine of a State-controlled economy' with the 'predatory Power State', whereby 'regionalism, self-determination and self-government become impossible'.[8] He asserted:

> Meanwhile the corpse of the Worcestershire which we know may linger on in a few small, scattered islands, preserved by public or private bodies as parks, or sites for holiday camps where, like children let out of school, the mass-minded helots of the Power State may go out to play, or to partake of their planned leisure.[9]

He went on to portray a bleak future of desertification, disease and famine.

Rolt's solutions were not ostensibly nostalgia, but 'self-sufficiency' and 'organic growth'. There was suspicion of 'scientific planners'. If an illustration of Matless' distinction between the 'planner-preservationist' and 'organicist' was called for, it could well be as follows:

The system cannot function unless industry and factory farm alike are under the control of the central bureaucratic machine. Despite all the neat maps and diagrams of the 'Town and Country Planners', none of the organic associations linking town or village with their rural environment can, under these circumstances, survive, while both, in their architecture and plan, will reflect this disintegration.[10]

Rolt did not provide much guidance as to how this disintegration might be halted, except his proposal that local museums ('those dreary repositories of irrelevant bequests') should be transformed into folk museums, exhibiting present-day tools and products of local industry alongside the past. He may well have been inspired by Joseph F. Parker and Alice Parker of Tickenhill Manor, Bewdley, who opened their Worcestershire Folk Museum on the eve of war in September 1939. This would ensure 'the feeling of continuity', but there he leaves it.[11]

Parker, chairman of British Rolling Mills Ltd of Tipton, had married into the Tangye family of industrialists.[12] After his wife, Alice – a former student of May Morris, William Morris' daughter – had researched the history of Bewdley, they had begun the collection of 'bygones' in 1933.[13]

Rolt commended the collection at Tickenhill, but contended that '"Folk museum" is an unhappy description which I use only because it is the accepted term ... I would prefer to call it a treasure-house'.[14] He described Bewdley as 'a typical example of an English country town at the close of the eighteenth century', but 'it was the product of the good workmanship of hereditary skill and knowledge of local materials, the flower of a way of life, now dead, and lying in state in the palace at Tickenhill'.[15]

Massingham had also encountered Tickenhill, 'the oddest and most bewildering heap of treasures I had ever seen [...] the attic-grave of old Bewdley's manifold industry'. He asked many questions about Parker, expressing some doubts about his industrial background but endorsing his proposal for a modern guild of handicrafts to revive the local watermill and the horn industry. Massingham recorded Parker's dream of 'drawing a charmed circle round Bewdley. Birmingham and all its age and works would be excluded from it and the town re-enriched with all of its former crafts and "misteries" with craftsmen and their apprentices to ply them.' Massingham asked Parker, 'Was the agriculture of the region to be restored with them to its former husbandry, since most of these trades rested upon it? This had not occurred to him but in no wise daunted him.'[16]

Part Two of Rolt's study, 'Place', demonstrates Rolt's admiration for craft. This was illustrated by his coverage of Harvington Hall, near Chaddesley Corbett, which had been purchased in 1923 from its owners – landowners from Coughton Court, Warwickshire – after decades of neglect. 'I am tempted to say that Harvington Hall is the most interesting historical domestic building in the western Midlands,' he asserted.[17] He had seen it in 1928, when much had been removed, including the staircase that was relocated to Coughton Court in 1910.

Rolt did not concur with the Victorian view of romantic ruins. 'To me it was a tragic sight,' he wrote.[18] After seeing it again in 1942, he praised a more sympathetic attitude to conservation, with the employment of a craftsman from nearby Elmley Lovett, who worked on Harvington between 1931 and his death in 1939. The rebuilding of a new staircase, which Rolt deemed a recreation, had been partly entrusted to the chair-maker from Droitwich, William Foulkes, who was much praised in *Worcestershire*. He did not comment, however, on the purpose of acquisition, which was to create a Roman Catholic place of pilgrimage that would pay homage to priests and others who had been persecuted. It had brought him into contact with the antiquarian Henry R. Hodgkinson, a solicitor practising in Birmingham. An early exponent of waterways history, Hodgkinson had written about the history of the hall, and would chair the Friends of Harvington Hall after retirement in 1953.[19] While the main restoration had been carried out by contractors, some work had been carried out by the Friends – an early instance of practical voluntary labour. He and Rolt must have become close, as his son Richard would accompany the Rolts on part of their Irish voyage in 1946.[20]

Rolt lamented the closure of the mill at Fladbury, on the Worcestershire Avon, 'a product of a logical and thrifty marriage between natural and human handiwork [...] yet this beauty at Fladbury is flawed, for the millrace is dry [...] it is one of the snapped links of a broken economy'.[21]

He made a rare excursion into the urban areas of parts of the Black Country, some of which were then within the boundaries of Worcestershire.

From Wrens Nest Hill in Dudley, he noted:

The furnaces no longer flare, for the iron trade is now concentrated in a few large works where only a shimmer in the sky above the cone-capped blast furnaces betrays their molten hearts. Though there are still miles of blackened wastes, they no longer smoulder and flicker with 'gob' fires, but lie cold and dead; an unearthly desert of cinders as sterile as Sahara, littered like a battlefield with the detritus of ruined walls and rusty machinery, and pitted with pools of still, dark water.[22]

He recorded the large modern factories, using electrical power, that had replaced these works with 'a mechanical inhumanity'. However, despite his stress upon ecology, he did not comment on how the severe air, water and land pollution brought about by the past industry was partly mitigated by more modern and concentrated processes.

He attributed this concentration to changes in trade and noted the movements of population that had crowded out the Bromsgrove nail trade a century before. Unlike some later Greens, he dismissed Malthus' prognostications about population outstripping the means of subsistence, as in Bromsgrove, 'those means had been taken away from it'.[23]

He did, however, find some remnants of the nail trade there after some enquiry: he met a Mr Weaver, aged 73, who made nails in a shed on a part-time basis. In appearance, he reminded Rolt of the late George Tooley of Banbury. Rolt commended his 'serenity which is the imponderable but priceless reward of one who has asked and expected little from life except the right of sturdy independence'.[24] He did not comment on the lack of independence involved in Mr Weaver's part-time employment as a labourer. Stressing the poverty that Mr Weaver had suffered, and continued to suffer, Rolt carefully described the way in which this craftsman made 'brush nails', which machines could not make as effectively. It was clear that this trade was dying out.[25]

For all the emphasis on craft, it is clear that Rolt was then dealing with relics – as with the death of the last boatbuilder employed at Tardebigge, in 1945. There is no sense that any of the craft skills would develop or be passed on – in that regard, Parker's idea, although Massingham thought it would fail, was apt.

## To *The Inland Waterways of England*

Although *The Inland Waterways of England* was published shortly after *Worcestershire*, it was written much later – after the IWA had been founded – and was almost the last of Rolt's waterways books to appear, being released at the same time as the Market Harborough Rally. In many ways, it successfully provided Rolt's personal portrait of the waterways at the onset of nationalisation and it remains a valuable document for that reason.

Its origins lay in the Inland Waterways Exhibition at Heal's in 1947, which brought the commission for *The Inland Waterways of England*. Not published until 1950, it was largely written in 1948. It was not based on archival research but provided a synthesis combining reading with much observation. Thomas Willan's work clearly influenced the opening chapter on river navigations – Charles Hadfield had lent Rolt his own copy – and some enthusiastic readers of *Narrow Boat* had responded with some limited research materials.

Rolt placed less emphasis on craft skills in his chapters on waterways construction; indeed, his coverage of engineers in his later biographies would not draw clear distinctions between those who developed professions, like John Smeaton or Thomas Telford, and an ingenious millwright like James Brindley.

He came closer to craft in his portrayals of maintenance work. This included his influential encounters with workers at Tardebigge, which had been well described in *Worcestershire*.

Of the blacksmith, boatbuilder, mechanic and lock-gate maker at the yard there, he commented, 'Except for the fact that they are employees of the same company and that their work is often co-ordinated and interlocked [these] might well be independent craftsmen'.[26] He did not consider the wage relation that defined their work, however much autonomy they might have. This failure to consider the relations of production was an omission that applied to much of his musings about a purported harmonious organic society and economy in the past.

In *The Inland Waterways of England* he focused upon George Bate, the carpenter who made lock gates for the Worcester & Birmingham Canal's many locks. This involved 'woodworking on the grand scale', something that was becoming increasingly rare. He suggested a continuity with preindustrial skills:

> But the shade of the mediaeval craftsman who hewed the timbers of the tithe barn or the hammer beams of the village church would feel at home in this carpenter's shop. The sight of the massive squared timbers of oak shaped and cut to tenon and mortice would occasion him no surprise, while the tools of the lock-gate maker, shell auger, adze and plough, would come sweetly to his hand.[27]

He felt that the recent innovation of metal lock gates could make these skills obsolete, while the shell auger was now 'almost a museum bygone'.[28] He turned to practical concerns for the future of waterways, contrasting the problems of a damaged wooden gate that was fairly easily mended with those of steel gates that could be distorted permanently and would need to be replaced.

By now, Rolt had become more concerned with practical measures that could enable waterways to become or remain viable, including the conservation of essential skills. He identified a lack of dredging as a major concern and was less disposed to praise traditional dredging like that using 'spoon' dredgers, 'which, by dint of a great deal of laborious effort slowly and literally spoon out the mud half a hundredweight at a time'.[29] Instead, he advocated that mechanical means, like dragline dredging, should be applied. 'Lack of capital and enterprise', which would have swept away 'craft' approaches to dredging, were identified as the problems, not some crass dismissal of valid craft approaches. He also advocated the steel piling of banks, which were much more prone to erosion by mechanically propelled vessels than by horse-drawn ones.[30]

His concerns here were less for craft skills than for the waterways to remain navigable for carrying boats to reasonable depths. This places him as a supporter of intermediate technology, fostering small-scale innovations to enhance older technologies.

He also laid stress on the skills involved in navigation, drawing a distinction between those developed for river navigation and those of the Midlands canals. He emphasised boats upon which families lived, although these were very much a minority.

Much comment was devoted to the traditional painting of some narrow boats of the English Midlands – those with which he was the most familiar, from his time in Banbury. He traced back the 'unique and striking' narrow boat to 'those days before the rise of the Puritan and the merchant when man expressed his joy in an unselfconscious richness of colour, colour which once gave rise to every rib, vault and boss of the grey Gothic cathedral a brilliance which we would soon find overwhelming'.[31] More recently, doubt has been cast both on his discussions of narrow-boat painting and the origins of boat people, and these need not be developed here.[32] He does not seem to have been generally interested in popular art.

He suggested Roma origins for the earliest boat people, citing the 'Gypsies' of Trafford Moss (and other Mosses near Manchester), who might have been attracted to work on the earliest Bridgewater Canal and then diffused their influence throughout Midlands canals that used family narrow boats. This may have reflected some elements of the 'diffusionism' of the 1920s, postulating the diffusion of agricultural practice from Egypt and Mesopotamia. This reflected Massingham's influence ('a comprehensive view of world-history', no less) from anthropological studies in the 1920s.[33]

He felt that the layouts of the narrow-boat cabin were similar to the Romani wagon: 'the layout of the narrow boat cabin corresponded exactly with that of the average gypsy wagon.'[34] Describing it in detail, he asserted that the cabin was 'an arrangement quite unique on the water because it was not derived from any marine tradition but was an adaptation of an ingenious plan for living in the minimum of space originally evolved and perfect by the Romani'.[35]

As for decoration, while 'they undoubtedly betray the gypsy love of bright colours', he conceded that forms of decoration like 'roses and castles' were unique to narrow boats. Yet he suggested connections between the Roma of Trafford Moss and the carts of the Balkan peasantry, which Roma from the Balkans might have brought with them.[36] 'The possibility may not be too fanciful,' he asserted, but it was based on guesswork.[37]

The traditions derived, he felt, from the family boat, 'and if the family boat is to be banished forever from our waterways, then that life and tradition will perish with them'.[38] He discussed the problems of conserving that way of life. Perhaps this could begin by making it better known what that life was like, but this perception was doubtful, as 'the canal boatman is the supreme example of the "private man". Proud, independent, resourceful, courteous he is the nearest surviving counterpart of that "bold peasantry" whom Cobbett championed.'[39]

He made practical suggestions about the enlargement of living space and improvements to payment systems and rationing, but he felt that illiteracy and education needed to be addressed. Illiteracy, which could be minimised by 'better educational facilities at turn-round points', was the main concern, 'for the rest the life of the boats, teaching independence, self-reliance and resource at an early age, provides a better education than any textbook as the children themselves bear witness'.[40] Against this, he placed a caricature: 'State bureaucracy can only function by imposing its own standard pattern of life upon the community, and by eliminating all those who cannot or will not conform to this pattern. Under such a dispensation the life of the canals seems unlikely to survive much longer.' He invoked the image of the small, defenceless and divided community against the all-powerful 'servile state', citing G.K. Chesterton and Hilaire Belloc.[41]

Noting the deserted canal-side inns, such as Tunnel House on the Thames & Severn Canal, he went further:

> They are the symbol of a whole world, a very ancient world, which is being submerged by our machine 'civilization' [...] It is a simpler world than ours; it contains more physical hardship, and because its vices, like its pleasures, are simple and overt, the puritan and the reformer have always been its embittered enemies.

Tunnel House would be restored, in part after a fire in 1952, but would close in 2020 over the more prosaic issue of the inability of the tenants to renew their lease on terms satisfactory to their landlord.[42]

Rolt did not suggest that men should be the slaves of tradition, but that a man who scorned tradition is the poorer, as 'Not only does he lose a precious sense of continuity with the pride in craft which this brings, but he is the poorer in the practical sense that he rejects methods of living or working which many generations have perfected'.[43]

It was not just craft, though, but areas of secret, private knowledge held by people that he could characterise as 'folk'. This is illustrated by the legend that clay linings would be puddled by driving a herd of cattle along the bed. While further details are discussed on page 158, this kind of legend reflects his feeling that many precious but obscure things and their meanings were being lost, so that he must at least record them in a form of 'preservation by record'.

Navigational skills were stressed in estuarial and non-tidal rivers. This involved intimate knowledge with 'accumulated experience of generations' and piloting skills in times of flood. He noted major differences between river barges and narrow boats – lack of decoration for the former and varied terminologies. Narrow boating involved quite different skills, especially when horses were used.

He noted the 'folk' elements in the lives of those who 'lived-in' on narrow boats. An evening in The Struggler in Banbury with boatmen and Roma travellers was recalled. Boatmen were noted for singing, but mostly songs from the Victorian music hall and nothing since the jazz era. Despite his enthusiasm for communal singing in waterways-related pubs, he did not seem to display an interest in music that could be described as 'folk music'.

Both boat people and the Roma engaged in step-dancing, and then one of the Roma told a story, 'a unique demonstration of the ancient art of story-telling as it was practised throughout the centuries which preceded the written word'.[44] He viewed this as the 'spontaneous art' of nomadic peoples, the foundation of 'great art'. He detected connections between the Elizabethan audiences who had first watched Shakespeare's plays and those capable of entertaining one another, without the influence of print or broadcast media, in the 1940s.

Watching the Roma's performance, he 'realized that here was the display of an art that had given birth to the first strolling player, and through him to all the greatness of the English dramatic tradition'. He drew pessimistic conclusions about this. If 'all great art has been built upon the foundation of this spontaneous art of the people [...] Had it not existed in Elizabethan England there would have been no Shakespeare', then 'the machine-minder for all his State education prefers the narcotics of the cinema', much would be lost.[45] This stress on the endangered popular roots of art is one that could be contested, but there would be echoes in the reactions against the strictures of F.R. Leavis, to whom only elites could appreciate the arts.

*The Inland Waterways of England* contained an elegy of sorts, as Rolt seemed to recognise that the world of living-in was coming to an end:

> If the family boat is destined to vanish from the waterways, then I am uncertain that the canal boatman will go also [...] If I am fated to see this happen, then at least I can console myself with the thought that I was not born too late; that I have known men to whom the canals and the boats were not just another job but a way of life.[46]

One feature of this is that there is no concession to the boats, crews, skills and ways of life on larger waterways in which the future of waterways transport was to lie. In this sense, *The Inland Waterways of England* was very much in the vein of *Narrow Boat*, his first book. It is doubtful that the canal-boat people whom he celebrated were part of any survival from ancient societies or practices rather than the remnants of practices that had been developed in the Victorian era, like the music-hall songs.

The emphasis on craft skills and their transmission other than through school instruction and qualifications begged the question of how these might be transmitted to future generations from the older men who still retained them. Parker's idea of a guild or the educational changes suggested in *High Horse Riderless* might indicate possible solutions. Later volunteers would need to develop the skills to work on restorations, while paid maintenance workers would have to be trained in specific skills.

Rolt himself would soon encounter the need for voluntary assistance in the restoration and operation of the Talyllyn Railway. There was little discussion of how people, often from white-collar backgrounds, would manage to carry out satisfactory work, although the early involvement of some professional railway staff must have helped. Other craft skills would be required to build new trading and pleasure boats and the conversion of older vessels for leisure use. Rolt, seemingly self-taught, had developed some of these for the reconversion of *Cressy*. Their bespoke and labour-intensive nature accounts for the high cost of pleasure boats against, say, mass-produced motor cars, or for that matter, trading boats against road-freight vehicles. Craft comes at a high price.

INTERLUDE
# LLANTHONY

Although the border market town of Hay was within walking distance of his home, Rolt's autobiography made little reference to it, bar visiting and being delayed at the railway station. The small suburban estate on which he lived was more notable for the nearby Cusop Dingle, which he explored in infancy with his nurse.

He noted later that his family were relatively unaffected by wartime shortages – his father's sports of shooting and fishing brought game and fish to their larder – and, indeed, proved little affected by the Great War. The young Tom also became familiar with a local farm, Lidiart-y-Wain, where he helped with the harvest. He recalled its reliance on horses and associated skills until the appearance of a Fordson tractor around 1917. He would develop an appreciation for non-mechanised farming, and especially for agricultural craftspeople, after the family moved to Stanley Pontlarge in rural Gloucestershire, in 1921.

While his home and the farm were under the shadow of the Black Mountains, these were not visited until June 1918 or 1919.[47] This proved to be a key experience, after his father hired a trap to take a picnic party over Gospel Pass (*Bwlch yr Efengyl*). A rough trackway climbed to the pass and another was followed down the Vale of Ewyas – 'so lonely, so secret and so strange', past the 'abbey', now deserted, at Capel-y-ffin, to Llanthony Priory.[48] Other family picnics would follow, but it was this first visit that made such a strong impression.[49]

'No building in Britain has so majestical a setting as Llanthony,' he asserted. On his first visit, however, he found the deserted abbey, which he termed the 'Monastery', 'a sad, depressing place'.[50]

His perceptions of the natural and historic environments came together at the priory. This medieval ruin surrounded by hills, with the sounds, smells and sights of both wildlife and traditional farming, combined and linked both his contemporary and future concerns. As he wrote later, 'Everything seemed to conspire to charm my five senses [...] its beauty has never failed to equal my expectation; and in recollection it was to prove an unfailing source of spiritual solace to me from that day to this'.[51]

He referred to the 'strange exaltation that I can only call a sudden awareness of the numinous, of the recognition of some profound reality behind appearances, that certain mountain country is capable of awakening within me'.[52] This sudden awareness seems also to have applied in Watergate Row in Chester, and it particularly applied to Llanthony, where he would spend many holidays for the rest of his life, exploring the surrounding area.[53] In its remoteness and antiquity, albeit one whose form had been much altered, including the opening of the Llanthony Priory Hotel in the former prior's lodgings within its boundaries, Rolt later felt an air of the mystical. Looking out of his hotel window one night in wartime:

> The summer night was perfectly still and calm, and a full moon had just risen above the dark protective wall of Hatterall hill. The clipped turf below was lightly covered with dew, and upon it the nave columns and arches cast shadows almost as dark and substantial as they themselves appeared. In this pattern of substance and shadow, the aspirations and the craftsmanship of long forgotten men, the loveliness of the landscape and celestial beauty of the night seem to have become inseparable parts of one whole so majestical that no words of mine can describe it.[54]

To Rolt, Llanthony seemed to represent the virtues of a preindustrial life. The priory was a partly reconstructed ruin; sold after the Dissolution, much of it had been steadily falling down. However, Tom was inspired by the setting and the spiritual basis of the Augustinian order, which had all too briefly occupied it. Rolt came to view the Middle Ages as a period when an organic social order corresponded to a harmonically ordered environment in which everyone knew their assigned place before industrialisation delivered the degradation of work and the despoliation of environments and landscapes. He reflected this in his novel *Winterstoke* (1954), in which the Cistercians are seen to rise again at the end against an industrial civilisation. He would perceive the priory as the embodiment of values that merited revival and conservation as much as the remains of the building itself.

By the time he had revisited this place that had so influenced him in childhood, he had become aware that Llanthony, as with other places, could be under threat. During his first revisit in September 1940, he discovered that the hillside of the adjacent valley, Gwyne Fawr, had been cleared, fenced off and planted with conifers – what he termed a 'vegetable concentration camp'. This reads somewhat insensitively, given the large number of humans who perished in the Nazi concentration and extermination camps, but it also seems to have reflected feelings about generalised human greed and folly, rather than any structural factors.

He would return from 1956 onwards, finding solace again; although, in his final volume of autobiography he complained that since the surfacing of the road over to Hay in 1955–56 and the opening of the Severn Bridge in 1966, increased numbers of visitors could make Llanthony a place to be avoided at times.[55]

Rolt's autobiography does not mention the fierce reaction of the diarist Francis Kilvert when he found other visitors, dismissed as tourists, who delayed his dinner at the Abbey Tavern, as he called it. Perhaps he might have sympathised, given his concerns that too many visitors, maybe of the wrong type, were being attracted. *Kilvert's Diary* was published as three volumes between 1938 and 1940, and Rolt felt that in retrospect it portrayed the country around Clyro and Hay in a way that suggested that little had changed between Kilvert's period of 1870–79 and Rolt's own of 1914–21.[56] However, this view was retrospective. It may not just have confirmed impressions but provided them, as he reconstructed the scenes and view of his childhood.

A similar sense of isolation can be felt today if the priory is approached from Hay over Gospel Pass, the highest such pass in Wales, on a lane that emerges onto open moor before descending through hedges past the hamlet and 'abbey' at Capel-y-ffin. This may be tempered by the great need to exercise care on a single-track road, which can become busy with motor traffic. Writing in the late 1940s, H.J. Massingham opened his *The Southern Marches* (1952) with a description of the Gospel Pass, accessed by a 'stony mule-track'.[57]

From the other direction, a much wider road diverges at the Skirrid Inn, twisting round the lower Vale of Ewyas, which opens into a bowl with the priory ruins in the centre. There has been further conservation work since Rolt's time, but the priory site is like many similar medieval church buildings in the guardianship of Cadw or English Heritage – existing structures stabilised along with interpretation boards and carefully mown grass. The area is recognisable from Rolt's descriptions, but is now much busier, since the Brecon Beacons National Park (now *Bannau Brycheiniog*) was designated in 1957, with the Offa's Dyke long-distance footpath completed nearby in 1971. The existing Half Moon Inn, just beyond the entrance to the priory, has added a campsite and bunkhouse.

The valley above Llanthony was threatened in 1951 with flooding to form a new reservoir, something that Rolt does not mention; although, after childhood, he had become aware that modernisation of the built environment could involve changes that would threaten places that he valued, like Llanthony.[58]

Oddly, although Massingham knew Rolt well by the late 1940s, he did not mention him in his book *The Southern Marches*. He recalled many visits, but 'in spite of its magical environment (and it appears from the bend of the vale to be surrounded by mountains), it seems to me almost entirely lacking in atmosphere'.[59]

Rolt felt that there was something implicitly spiritual about some connections with the past, and explicitly spiritual about the priory in its rural location. The latter was reinforced by the feeling that the mystic writers Thomas Traherne and Henry Vaughan had been influenced by the same Black Mountains. The building, its history and the mountain setting led him to perceive that 'I must have mountains around me [...] I am aware of [...] that strange exaltation that I can only call a sudden awareness of the numinous, of the recognition of some profound reality behind appearances, that certain mountain country is capable of awakening within me'.[60] Something in the Talyllyn Railway's setting may have appealed to him, but despite his affinity with canals and railways, it is a visit to Llanthony Priory that should convey one of the most significant influences on Tom Rolt.

Forestry in the upper valley below Hatterall Ridge. Rolt decried this kind of planting by the Forestry Commission as a 'vegetable concentration camp'.

A distant view of Llanthony Priory in 2010 from the north, with an incongruous corrugated-iron shed in the foreground.

The Priory entrance, with the author's wife, Sara, in the gateway. (Author photograph, by permission of Cadw)

The Llanthony Priory Hotel in 2009. (Author photograph, by permission of Cadw)

The north side of the ruins. The view looking south through the priory ruins perhaps reflects the sense of enclosure that the surrounding hills and mountains provide. (Author photographs, by permission of Cadw)

# 7

# ENCOUNTERS WITH TRANSPORT HISTORY: WILLAN, HADFIELD AND ROLT

Tom Rolt would come to be regarded by many as an historian of transport and industry, and his writings and involvements would inspire many into an appreciation of history through various channels.[1] However, while until the mid-1950s he wrote about history, this was not rooted in engagement with formal historical research. He would later record that *Red for Danger* (1955) was 'the first book I had tackled which entailed research' and his later Brunel biography involved much 'detective work'.[2] This confirms Charles Hadfield's assertion that, although a close friend, Rolt was not really an historian of waterways or railways and he rarely had time to carry out detailed formal research.[3] This may well have been the case for much of Rolt's earliest writings, so that their continuing value lies elsewhere. Nevertheless, he did carry out extensive research in ways that diverged from the models of research and writing to which many transport, industrial or economic historians tended to conform.

This chapter will examine two near contemporaries of Rolt and their approaches to waterways history, alongside Rolt's contrasting, if complementary, approach to the historical. These were Thomas Stuart Willan (1910–94) and Charles Hadfield (1909–96). The coverage of their lives is uneven, as the lives and work of Rolt and Hadfield have been much considered, by myself in the case of Charles Hadfield.[4]

All three were 'outsiders': Willan (no relation to Kyrle Willans) was from Hawes, in the Yorkshire Dales, Hadfield from South Africa and rural Devon; and Rolt from border England. All three had distinguished careers and all contributed to waterways and other history. But in what ways?

## Willan: Academic, Economic History and Transport

Thomas Stuart Willan was little known outside his academic field. He seems to have spent his whole working life in academia, publishing *River Navigation in England* (1936), *The Coastal Trade* (1938) and various detailed monographs on individual river navigations, the last of which were in 1965.[5] He lectured in economic history at the University of Manchester from 1935 until retirement in 1973, having been made professor there in 1961.[6] Unlike some academic contemporaries – T.S. Ashton, for instance – he does not seem to have become much involved in public affairs.[7]

Willan was born in the Yorkshire Dales. He would later recall horse-drawn transport in Yorkshire and Westmorland, and he retained family connections in Aysgarth and Hawes.[8] Hawes lay at one of the furthest points in Britain served by river or coastal navigation, and perhaps this distance was one influence. After his PhD thesis at Oxford was conferred in 1935, it was published in two volumes: on river navigation and coastal trade. He insisted that these two subjects were closely related and only divided for ease of publication.

Willan's two books from the mid-1930s proved formative. Prior to this, the view had descended from Thomas Macaulay onwards that neither inland nor coastal water transport was of much significance before the canal age. The emergence of the academic discipline of economic history in Britain seems to have fostered some interest in water transport's role in economic development. It was a relatively new discipline – the Economic History Society, with an international membership, had been formed only in 1926 – coloured partly by some of its leading participants and their encounters with industry and employment. Some appreciation of early leading figures may provide some insights.

Professor George Unwin (1870–1925) would become Britain's first professor of economic history, at Manchester. Growing up in Stockport, he worked from 1883 in a local hat manufactory as an office boy, learning bookkeeping. After studying at night school and then Cardiff University College and Lincoln College, Oxford, in 1898 he went to study economics in Germany – academic economic history there was more advanced than in Britain, laying much stress on documentary research. After he went to work as secretary to Lord Courteney, a former Liberal MP, this provided time to pursue research in the British Museum and the Public Records Office, investigating voluntary organisations.

Unwin became lecturer in economic history at the University of Edinburgh in 1908 and in 1910 was appointed as the first chair in economic history at Manchester University.[9] His practical experience working in a factory may well have encouraged him to investigate industrial history. While his earliest archival research investigated medieval guilds, in 1913 many records of the Manchester cotton manufacturers McConnel & Kennedy were discovered for the period from 1795 to 1835. His colleague, G.W. Daniels, used these records in *The Early English Cotton Industry* (1920), for which Unwin wrote an introduction.[10]

In 1920, a large store of papers, discovered in a derelict mill at Mellor, near Marple, were rescued for the university by a team from Manchester. Unwin used these to write *Samuel Oldknow and the Arkwrights: The Industrial Revolution at Stockport and Marple* (1924).[11] Corley claimed that this was 'the first British business history to be based on a comprehensive use of corporate archives'.[12] He contrasted this with 'the superficial and laudatory accounts from the pens of retired company secretaries or family members'. Rolt's later commissioned company histories might fall between these categories.

Unwin, a friend since 1912, prompted Thomas S. Ashton to return to Manchester in 1921. Ashton, who would develop a view that the Industrial Revolution had been beneficial to workers, also favoured archive-based work. The use of archives and industrial economic history was unusual at the time, but this would partly form a base for some post-war British transport history work.

Willan's Oxford supervisor, George Clark (1890–1979), does not seem to have been a major influence, although Willan followed a similar approach by tracing details through archive sources in preference to the use and development of economic theory.[13] *River Navigation* contains only two references to Clark, one of which points out his mistaken assumption that there were no pound locks on the Thames.[14]

Willan was directly influenced by Canadian academic William T. Jackman's study of transport history, which covered the period from 1500 to the nineteenth century. Although Jackman (1871–1951) relied heavily on legal and statutory documentation, he did develop some insights into economic factors.[15] This study drew praise from John H. Clapham: 'Mr. Jackman has little competition to face in the history of transport by water, and it is to this that he makes his most important contributions.'[16] Clapham (1873–1946), Professor of Economic History at Cambridge from 1928, would prove more influential upon Willan, especially in his emphasis on the empirical and upon avoiding broad generalisations from limited materials, unlike Rolt, perhaps. However, Clapham's *Economic History* (1929), beginning in the early railway age, offered little comment on river or coastal navigation, while his posthumous *A Concise Economic History of Britain* (1949) did not mention Jackman's or Willan's work.[17]

Willan's work was cast very much in the vein of economic history, rather than any separate fields of transport or industrial history (if these could be said to exist before the 1940s). His works do not suggest that he actually visited waterways, ports or other forms of transport. The principal sources were very much academic or antiquarian, employing archival research. His conclusions were mostly economic, arguing that water transport was much more important in the pre-canal period than had been thought.

There were minor errors in *River Navigation in England*. For instance, the River Dane in Cheshire was never made navigable, and there were confusions between the sizes and types of river and canal craft, especially narrow boats. Willan himself corrected several minor errors in his 1951 monograph on the Weaver, which covered the whole eighteenth century.[18]

Willan also studied coastal trade and in *The Inland Trade* (1976) returned to this and to river navigations and road transport in the sixteenth century, a period before that covered in *River Navigation*.[19] Stressing the importance of all three, he was acid about sources like the eighteenth-century agricultural writer Arthur Young:[20]

> Perhaps historians of transport, with their outlook of suburban motorists, are too concerned with the conditions of the road and too obsessed with Arthur Young's ruts four feet deep (which must have been made by wheels about nine feet in diameter).[21]

Willan was scathing about much post-war writing on railway history. For instance, in 1964 he reviewed H.P. White's *Greater London* volume in the David & Charles *Regional Railways* series in these terms:

> Clearly the book is intended primarily for the railway enthusiast, or perhaps for the railway enthusiast who is also a Londoner. He should find satisfaction in the wealth of detail, the pleasant pictures and the excellent maps. Others may well regret that so much energy is now devoted to this tedious and unrewarding form of railway history.[22]

However, Willan was more impressed by Hadfield's work. He commented about the second edition of *British Canals* (1959):

> His book is clearly based on the best secondary authorities and on the records of the canal companies themselves [...] Its virtues stem partly from the author's deep knowledge and partly from his approach to his subject. His knowledge enables him to deal with aspects of his subject which are often neglected. Thus he stresses the continued importance of river navigation during the canal era, which is frequently forgotten.

The latter comment demonstrates Willan's awareness that the significance of river navigation had continued beyond the historical period that he had covered. However, he went on to denounce railway and canal enthusiasts, citing:

> ... that arid antiquarianism which afflicts some historians of transport, especially perhaps the railway fanatics. Though Mr Hadfield both knows and loves his canals, his approach to their history is objective ... Here is no sentimental yearning for the obsolete which afflicts so many canal enthusiasts, who ascribe the decline of the canals to the unfair competition of the railways.[23]

Although it is not clear whether Willan was aware of Rolt's work in 1959, these comments might apply to some whom Rolt had inspired rather than to Hadfield's more hard-headed approach. Hadfield was, in fact, one of the few historians of waterways who would be directly inspired by Willan.

## Hadfield: The Making of History and Types of Transport History

Ellis Charles Raymond Hadfield (1909–95) grew up in South Africa, but after moving to rural Devon in 1923, carried out schoolboy, but still serious, research using solicitors' papers relating to the Grand Western Canal, the only local waterway. The extensive reading of files of the *Exeter Flying Post*, whose coverage reflected much in regional and national history beyond waterways, seems to have influenced his later archival research.[24]

Hadfield found few books in print about waterways in the 1930s but was probably influenced by railway company histories like E.T. MacDermot's *Great Western Railway*, which made some use of archival sources.[25] His studies into economics at Oxford must have engendered some familiarity with the growing field of economic history. After reading *River Navigation in England*, Hadfield contacted Willan, who advised him about the availability of pamphlets, plans and papers in the Institution of Civil Engineers' library. The two met at the annual meeting of the Economic History Society in 1939, but Hadfield felt daunted by the academic tone of the meeting and took no further part.[26]

In the 1940s, Hadfield was much involved in politics and trade unionism in London. While he later maintained that his earliest canal books were not written for didactic or scholarly purpose, the first edition of *British Canals* (1950) provided an excellent summary of what was then known and incorporated some archival research. A subsequent series of books attempted to detail the history of every waterway in England and Wales, using the major source of archive materials in the British Transport Historical Records, which had been available since 1951. Their historical coverage tended to be much lighter for the twentieth century, partly because archival sources were sketchier and less often available under what was then the fifty-year rule. Additionally, Hadfield had become involved in the making of history and sought to avoid controversy.[27]

As was stressed in *Charles Hadfield: Canal Man and More*, Hadfield's involvement with the making of post-war waterways history was multifaceted. He helped to found the IWA in 1946 and the Railway & Canal Historical Society in 1954, after which he organised numerous visits to waterways to increase his own knowledge and that of others, partly seeking to encourage research into the historical.

In this manner, his approach differed from that of Willan. Hadfield did not attend academic conferences or write much for academic journals after his first publications in *The Economic History Review*, and some academics viewed work like his as merely that of an enthusiast. Examples of this, in which his detailed histories were never acknowledged, even in footnotes, were cited in *Charles Hadfield: Canal Man and More*.[28] Against this can be set academic praise from the editor of *The Journal of Transport History* in 1987, while Hadfield also researched and wrote *William Jessop, Engineer* with Professor Alec Skempton (1914–2001).[29] Skempton, with a background in practical civil engineering and later editor of the

*Biographical Dictionary of Civil Engineers* (2002), had investigated early river navigation engineers in 1953 with Willan's help.[30] He was introduced to Hadfield through Rolt, who had mutual connections with the Newcomen Society. Skempton was adamant that Hadfield's work, although based on less technical knowledge, was highly regarded.[31]

Hadfield's involvement with David & Charles publishers from 1960 enabled him to consider both short, popular pamphlets by enthusiasts, and weightier volumes of history by professionals and sometimes those with academic backgrounds. The encouragement of publication brought forward many new authors and fostered much popular interest in transport history; by way of contrast, Willan's detailed studies were published by local learned societies with a limited circulation and influence.[32]

## Rolt and the History of Transport

Like Hadfield, but unlike Willan, Rolt's interest in history was inspired by direct involvement with waterways and other transport. While he helped to popularise interest in transport history, this was in very different ways to that of Hadfield. As has already been stressed, he was inspired by landscape and the changing rural and industrial world.[33] He also found Samuel Smiles' engineering biographies inspiring, stating later, 'He had a greater influence on me than almost any other writer. I think it very significant that I was never introduced to him or to his subject while I was at school.'[34]

W.A. McCutcheon, an early investigator of industrial archaeology and author of *The Canals of the North of Ireland* (1965), made a helpful contrast between Hadfield and Rolt's approaches to history. Reviewing Rolt's *Navigable Waterways* (1969), he contrasted:

> Charles Hadfield, whose carefully researched regional histories of British canals are the standard works in the field, and the present author, whose elegant and lucid style makes light work of the presentation and analysis of a wide range of factual information. No one is better qualified than Rolt to explain the general development of inland navigation in Britain, or to describe at length the highways and byways of our canals and navigable waterways.[35]

A further contrast would be in the breadth of Rolt's later studies, from waterways to biographies of engineers, and from road transport, railways and industrial archaeology to business history.[36]

Rolt's observations and deep practical knowledge enabled him to write from a familiarity with waterways and other transport structures, relating much history to features that might still be visited and observed. Influences included Smiles, alongside a diverse set of sources. Formal historians, especially academics, seem to have been much less influential. Beyond literary sources and factual ones, such as H.R. de Salis' *A Chronology of Inland Navigation*, in many ways, Rolt relied on personal observation: an acute feeling for landscape, a strong practical comprehension of technique (especially involving craftspeople), and an ability to converse with boat people and other workers and to absorb their knowledge into his.[37] One consequence was a strong 'folk' element from these encounters.

He was unimpressed by academic economic historians, referring to 'a good deal of sniping between the true engineering historians […] and the economic historians with each side accusing the other of presenting a partial view of history'. He came down firmly against the economic historians, preferring 'the humanistic approach to engineering history. It shows us that the great engineers were great creators and not economic animals.'[38] Whether Willan was aware of Rolt's work and views is unclear.

Some aspects of Rolt's practical approach can be discerned in his 1940s research into the Leominster Canal – tracing remains on the ground, talking to local inhabitants and gleaning details from local books like R.C. Gaut's *Worcestershire*.[39] The results lay somewhere between professional and enthusiast history.

*The Inland Waterways of England* was published much later, when publicity for the waterways cause was having an effect. Much of its first chapter leant on Willan, while its second chapter, on the canal era, contained some misleading points from published sources, attributing the Rochdale Canal to Rennie, not Jessop, while the engineering of the Sankey by the Liverpool Dock engineer Henry Berry was attributed to John Eyes.[40]

These details did not mar the main chapters in the book, which presented a personal portrait of the already historical rather than a detailed history. Over seventy-five years later, this remains invaluable in portraying and describing such questions as the use of horses, approaches to waterways maintenance and boatbuilding. At this point, Rolt's approach was often intuitive, viewing what lay on the ground and frequently comprehending its engineering; similarly, he understood mechanics and the use of craft skills.

Accuracy in description and acute observation did not always match the archival evidence that later emerged. Rolt also lamented the modern world, and a detailed account of recent boatbuilding activity ended with the observation, 'It was sad to reflect that I might not see her like again'.[41] There was, by now, a campaigning factor, a plea for the conservation of valuable ways of life, although the means by which such conservation might be undertaken were not specified.[42]

Other observations diverged greatly from the kind of evidence-based history of Hadfield and Willan. Thus, Rolt commented on the impact of canals and railways upon landscape and attempts to mitigate the visual damage to the amenities of country estates. He felt that this had scarcely proved necessary for waterways, but:

> ... wherever man worked in co-operation with nature there could be no divorce between beauty and utility. But as soon as the relationship between man and nature changed from co-operation to conquest the two were divorced and the task of reconciliation became progressively more difficult.[43]

While canals had often made a positive contribution to landscape, and some railway engineering may have mitigated part of the worst damage to the landscape, 'even the railway builders were better off than our planners of today who attempt the impossible task of inducing the ferro-concrete factory, the dual track highway, the "pre-fab" housing estate, the pylon and the power station to conform with the English scene'.[44] This is a familiar theme from Rolt, and Matless would deem this kind of view 'the characteristic organicist stance of embattled melancholy'.[45]

Rolt's approach was, perhaps, aimed much more at those who would explore waterways, and indeed landscapes, for themselves; not only might evidence on the ground prove more important than that stored in archives, but he highlighted the appreciation of landscape and a perceived potential harmony of nature and engineering.

The role of Hadfield's research in determining the background to various obscure waterways was acknowledged, but the need for fieldwork to determine precisely what had been built was emphasised. Rolt asserted that there was no trace of the Aylestone Hill Tunnel at Hereford (on the Herefordshire & Gloucestershire Canal); it was uncertain what remained of the Chard Canal; and the remains of the Bude Canal had not been systematically recorded. All would, in fact, be investigated by Hadfield in the 1950s for *The Canals of Southern England* (1955) and *Introducing Canals* (1954).[46]

Such minor inaccuracies did not mar what was often very readable and well-crafted prose, which conveyed a strong sense of the waterways scene in the 1940s and drew in and inspired much popular interest. When Professor J.R. Harris (1923–97) wrote later, 'Rolt was a fine scholar, but not a professional academic', this was less a criticism than a tribute to his ability to enthuse those outside academic and professional circles.[47]

## Leominster Canal and Avon Explorations

To consider the varied kinds of history produced by these three authors, specific examples will be considered. All covered the Warwickshire/Worcestershire Avon and the people associated with waterways.

Little detail seems to have survived about Rolt's early methods of investigation. However, correspondence with Charles Hadfield provides some impressions of his investigation into the Leominster Canal in Worcestershire, along with their limitations.[48] He had encountered references to a 'Kington and Leominster Canal' in Gaut's book.[49]

While *Cressy* was moored at Tardebigge during the summer of 1945, he visited part of this canal, which had been closed since 1857. In the Newnham area, he found that 'In places the bed had practically disappeared but for most of its length it was intact, with over bridges, a short tunnel, and a fine stone aqueduct over the River Rea'.[50] Alerted by Hadfield that work on this uncompleted canal might have extended towards the Severn at Stourport, he visited Stourport to find no trace there.

He returned to the Mamble area by road on 12 August:

> We passed the sites of three locks (one with lock house) and then came to what had obviously been a wharf complete with a large brick house and stabling. A hundred yards further on there was an obvious 'winding hole', while a disused track (which might once have been a tramroad to judge from the careful grading) led away in the direction of the Mamble pits.[51]

These assumptions would prove largely accurate. Hadfield had told him about a reputed tunnel at Southnet. One 'old local' told Rolt that the west end of Southnet Tunnel had been blocked up many years before, but he found the east end (open to this day) himself. Over the planned long tunnel at Pensax, Rolt noted, 'Local information is scanty and contradictory',[52] but he recorded in *Worcestershire* that 'the canal ended ingloriously at the foot of rising ground on Southnet Farm near Mamble where all the efforts of the engineers to complete the 3,000-yard Pensax tunnel through the crumbling cornstone failed'.[53] This somewhat romantic picture would be dispelled by later research by Richard Dean, which showed that the very limited work on the tunnel ceased for financial reasons.[54]

For the rest of the canal, Tom also noted, inaccurately, 'I am told that it is possible to trace it all the way to Kington'.[55] He based this on a reading of Joseph Priestley's *Historical Account of the Navigable Rivers, Canals and Railways of Great Britain*, which Hadfield had lent him.[56]

Willan's period preceded canal development, but he did comment on the Lugg and Wye rivers in the same regions. His research involved private papers, so he inevitably considered individuals, especially the promoters and controllers of river navigations. He explored conflicts and connections between the various Wye and Lugg undertakers and their interest in receiving compensation for the removal of weirs that they owned. The difficulties in discovering much about river engineers and surveyors were stressed.

Further down the social scale, his coverage of boat people was vivid but detached. He pointed out the dangers involved in bow-hauling, and its casual and arduous nature, whereby many boys lost their lives, but:

> It is dangerous to attempt to judge the character of a class of men who lived two or three centuries ago. If the present can look impartially on the past, the past could not look impartially upon itself.[57]

He went on to cite evidence of the misbehaviour of some boat people and the honesty of others. Willan was very wary of generalising and stressed the very limited evidence about the standard of living of these workers. In the post-war period there would be debates in British academic history over the nature of the Industrial Revolution and its impact on workers and their standard of living.[58] While Willan's work showed that the technology for later canal development was already developed by 1750, his work mainly adds uncertainty to debates about the standard of living.

Hadfield, in *The Canals of South Wales and the Border* (1960), produced a more accurate account using local newspapers and other local sources, such as minutes of the Shrewsbury and Hereford Railway Company, which acquired the canal. He recorded that the Southnet Tunnel had remained unused after part of it collapsed, and a tramroad (horse railway) may have been partly built in substitution for the canal line from the Severn to Southnet. He focused on law and finance: £12,000 was raised in 1858 to acquire a canal on which expenditure by 1801 was £93,500, while no dividend was ever paid to shareholders.

## The Early Warwickshire Avon Navigation

Accounts of the early history of the Warwickshire/Worcestershire Avon developed with greater knowledge and interpretation. Jackman had made a single reference to this river and its improvements by William Sandys, a Worcestershire landowner. Willan devoted substantial coverage to the Avon, drawing details from the Calendar of State Papers of a conflict between Sandys and one of the Worcestershire commissioners. He indicated that the class connections of opposition to navigation improvements might have been a factor in the English Civil War during the 1640s.[59] The involvement of Sandys and his relatives elsewhere was stressed.

While *River Navigation* contained no discussion of the engineering works on the river, Willan described these in a 1937 article in *The Economic History Review*.[60] He provided a more linear account, including discussion of the pound locks and navigation weirs, citing Habington's *A Survey of Worcestershire*.[61] He was dismissive of Andrew Yarranton's contemporary claim to have made the Avon navigable, and especially of Professor J.U. Nef (1899–1988), whose misleading statements in *The Rise of the British Coal Industry* (1932) included the citing of Priestley's references to the Bristol Avon.[62] His principal conclusion about the Severn and its tributaries was economic in focus: 'These rivers appear to have carried on their surface almost the entire trade of the district through which they flowed: their navigation was an indispensable factor in the commerce of the west.'[63]

Charles Hadfield repeated this emphasis on the Severn in *British Canals* (1950), providing a more detailed account in *Waterways to Stratford* (1962). This also quoted Habington and commented that it remained unclear what Sandys actually achieved. Detailing the involvement of

Yarranton, Hadfield speculated as to which locks were built; it was especially unclear which navigation weirs had been built on the upper section above Evesham. This account used sources in local records offices and the minutes of Stratford Corporation. Unusually for Hadfield, it included itineraries based on fieldwork to enable enthusiasts to visit key sites.[64]

Rolt's approach to the Avon was mostly derived from practical encounters, partly from his 1938 trip in *Miranda* and later visits. This was mainly reflected in his descriptions of the Avon in *Worcestershire* (1949). Following the line of the river downstream through riverside settlements and landscapes, Rolt asserted that the locks and navigation weirs were built by Sandys but provided few historic details. Worth recording, however, was his contemporary description of the navigation weir at Cropthorne:

> A wooden gate fitted with sluices is set in the weir, being similar to a lock-gate except that it necessarily has no balance beam but is swung by means of a winch on the river bank [...] Assuming the gate to be open and the river making a level, a craft heading upstream passes through the gate, closes it and then lies on the weir sill until the level of the reach between Cropthorne and Fladbury Lock has made up sufficiently.[65]

This level of detail would not have appeared in Willan's or Hadfield's work from experience, despite documentary research. Rolt's portrayal reflected his pessimistic interpretation of industrialisation, regretting much decay in the contemporary world. He lamented the closure of the mill at Fladbury, but it would be impossible to imagine Willan, whose work was firmly rooted in sources about the preindustrial period, engaging with such concerns. Hadfield was similarly unsentimental.

In *The Inland Waterways of England*, Rolt recorded that the two flash locks on the Avon were the last to survive and 'They are therefore of the greatest historical interest, and it is to be hoped that their preservation for many years to come can be assured'.[66] However, as part of the revival of navigation on the Avon, the last working gate was destroyed in 1956.

## People, Transport and Waterways

Hadfield and Rolt would not be involved in academic debates like those over working conditions. Oddly, given Hadfield's background in the labour movement, he did not develop any major coverage of the lives of those who built and maintained the waterways and worked on boats. His discussions of workers associated with waterways proved so limited that I enlarged these substantially in the second edition of *British Canals*. This was a major criticism by Professor Ashton of the first edition of *British Canals*, although he acknowledged the dearth of archival sources.[67] Along with the lack of accessible empirical evidence (despite the opening of British Transport Historical Records, which involved mostly company records), it seems that Hadfield tried to minimise discussions of contemporary matters with which he had been involved. Most of his historical coverage ended at 1947, when he began to influence the course of history; it was also, coincidentally, at the time of transport nationalisation.[68]

When it came to the canal engineer Thomas Telford, even Rolt would concede that his personal life was an enigma, referring to:

… a facade behind which few, if any, of his contemporaries were ever permitted to pass […] that the man who devoted so much of his leisure to reading and writing poetry was a more complex personality than many of his friends supposed cannot be doubted. But he keeps his secret well.[69]

Hadfield had long been suspicious about Telford's precise involvement in works (primarily Pontcysyllte and the Caledonian Canal) with which both he and leading engineer William Jessop had been involved and published the results as *Thomas Telford's Temptation* (1993). In this, he speculated about Telford's private life and motives.

For once, he developed a modest conspiracy theory without reference to other historical conspiracies, and also made an analysis of successive texts by and about Telford without utilising the kind of poststructuralist deconstruction that all three authors would have roundly condemned! *Temptation* is a tribute to those whose revisions of historical accounts are based on re-examinations of sources and interpretations, and Hadfield's closing writing (literally the last words he published about waterways) asserts, 'Only in our own days, with the opening to students of so many more records and increased interest in inland waterway history, have efforts after truth begun to replace legend. This book is one of them.'[70]

Rolt focused on practical people, and thus engineers and workers, rather than promoters, shareholders and managers. He was, ironically for one who would later write so much about engineers, perhaps on surer ground in his portrayals of waterways workers and boat people; the latter had, in many ways, inspired his interest in waterways. He had met many boat people in Banbury between 1939 to 1940 (and post-war) and maintenance staff and carriers at Tardebigge. The chapters on maintenance and boat people in *The Inland Waterways of England* do not cover the history of associated workers in any chronological or anthropological sense but provide many details from the 1940s that reflect the past, including folklore.

Rolt only began formal archival research, with railways not canals, after his involvement with the latter had largely ceased. There is little material to back up his assertions about cultural survivals, bar his observations, although these could be followed up. Although he quoted the 1911 work *The Village Labourer* by Barbara and John Hammond, since the 1940s much social and labour history has developed, especially with the growth of social scientific and oral history research. His ability to paint portraits of contemporary workers was valuable, but the inferences about their past were more speculative and partial.

## Some Conclusions

Willan, Hadfield and Rolt were among the most significant individuals in the writing of post-war British waterways history. Willan developed a sound academic base for a hitherto little-known area of transport history and inspired others like Hadfield. Hadfield laid down the foundations for detailed factual histories of every British canal and river navigation, and played a critical part in securing the retention of waterways for leisure use. He also, through David & Charles and diverse other means, encouraged a generation of researchers into waterways history. Again, in diverse ways, Rolt enthused substantial numbers of people to explore research and become involved in waterways, railways and other forms of historic environments.

The kind of antiquarian work fostered by Willan (and Hadfield, to a lesser extent), based on the meticulous examination of sources, remains valuable and provides foundations for more speculative work. Where direct sources have left decided gaps in knowledge, as with the Avon, it may be that apparently unconnected sources, especially related to other forms of transport or to river management, may begin to produce a clearer view. Equally valid would be the development of academic approaches that build upon the factual foundations but draw wider conclusions.

Both Willan and Hadfield attempted to place waterways history within a broader sweep of transport and economic history. Willan was writing in a newly established field, but Hadfield's attempts to provide a general regional history of transport, beginning with *The Canals of Southern England*, were only partially successful. One deterrent lay in the problems of publication in fields in which many potential purchasers and readers were enthusiasts for detail.

The value of the encyclopaedic and comprehensive regional histories produced or edited by Charles Hadfield have not been diminished by time, and it is unfortunate that some academic commentators failed to acknowledge their validity. This reflects a tendency by some to draw boundaries to exclude non-academic researchers, perhaps fostered by the kind of enthusiast railway history work that Willan disdained.

Rolt's work illustrates and conveys the impact of a deepening personal and subjective sense of loss – the urge to record in words, photography and drawings that provide many valuable observations for later generations. The sometimes questionable interpretations – Hadfield's legends – behind that sense of loss or of what was being lost do not remove the acuteness of portraits that go beyond the factual. In a sense, Rolt is one of the pioneers of 'heritage', but rather than dismiss it as a falsification of history, we need to acknowledge its ability to draw in and develop interest in more serious work among contributors who may have been deterred by more conventional forms of history.

Rolt's focus on what might be considered 'folk' elements in experiences and practices is also suitable for much further work. His feeling for and celebration of craft and craftspeople presents a significant contrast with the more factual and archive-based histories of Hadfield and Willan, and might prove more compelling to a wider and less-specialised readership.

Others have taken up some of his concerns – over the painting of Midlands narrow boats, for instance – although different interpretations have emerged. The ways of life that he discerned have also been examined, although the extinction of living-in on narrow boats since the 1970s has made them difficult to research in depth.

Despite this, there is now a wealth of material in oral history interviews, much of it untapped, which requires interpretation. Oral history has developed since the time of Rolt's everyday conversations – which provided a dense texture to his observations – far from, and preceding, the tape recorder. Oral history material provides an emphasis upon experiences, feelings and dense personal information about everyday lives. Without it, there is much that can be permanently lost. Rolt caught this when he related a tradition that puddle clay was consolidated by driving herds of cattle along the canal bed and acknowledged that:

> while this is only hearsay, the legend is worthy of permanent record before, in company with many other little-known features of canal history, life and work, the recollection dies in some lonely cottage with an old lengthsman and the winds of this age of dissolution blow away the slender threads of folk memory.[71]

Thanks to Rolt's mediation, and that of others who have followed, some of this 'folk memory' has not been entirely blown away. But there is still much vernacular knowledge and witness that is unwittingly fading.[72]

All three authors were innovators: Willan in river navigation, Hadfield in detailed canal history, and Rolt with his blend of reading and observation. They also set out alternatives to previous approaches to and interpretations of history. There is scope for further reinterpretation, some based on new evidence and some on new forms of analysis, often re-examining older writing. To consider new approaches in pursuit of Hadfield's 'efforts after truth' is not to disdain the legacy of these authors.

One possible future direction for waterways and other transport history may be indicated by a theme in Rolt's work that he did not take further. He asserted that 'The whole history of inland transport has been characterized by man's rapidly increasing mastery over the obstacles imposed by the configuration of the landscape'.[73] He went on to commend regional cultures and architecture that 'has grown out of the ground upon which it stands'. While he compounded this with great regret about the uniformity imposed by modern development, future research could utilise a form of environmental history. This would be based not entirely upon the characteristics of local environments, but upon the cultural interactions involved in the provision, development, maintenance and operation of transport systems and their environmental impact. Such approaches could complement the cultural emphasis in much current academic transport history.

## INTERLUDE
# BANBURY

It may be best to begin with a walk through Banbury to the waterside at St Mary's Church – the church whose tower Rolt climbed in 1939. A path to the side and rear leads down into Church Walk and then into Church Lane. Turning left here, there is an older street pattern, which has been retained, but the lanes are pedestrianised and most buildings are modern replacements. At the end of Church Lane is Parsons Street, with Ye Olde Reine Deer Inn opposite.

Rolt would have disapproved of the 'Olde' language but approved of the Reindeer, as it was then called, where 'a magnificent pair of oak doors beneath the courtyard archway remain to speak eloquently of the past'.[74] He would on go to accuse brewers of damaging historic buildings both externally and internally, and putting 'money-making' in front of 'the poor man's parliament and platform, his playground, and his solace after labour'.[75] He might have added the closure of smaller, less remunerative pubs, but there is much truth in his assertion.

Parsons Street is also pedestrianised and opens out onto Market Place, now also pedestrianised. Street markets are still held here and Rolt felt that these saw Banbury at its best, although he favoured the weekday farmers' market over the more general Saturday one. He was scathing about the way in which shopfronts had disfigured historic buildings, and there remains some evidence of this in shops that flank Market Place.

When he first arrived in Banbury, Factory Street was a very narrow street that opened almost imperceptibly off the marketplace, in a district of mainly nineteenth-century buildings. These included Rolt's favourite pub, The Struggler, which was a small beer house on a corner plot; it closed in the early 1950s. This area, in which some boat people and carriers had lived, became very decrepit, and demolition was under way in 1973 to build a new shopping centre.[76] At the time, it was proposed to provide a marina for pleasure boats on the site of Tooley's Yard, although this was still very much operational, with Herbert Tooley living on site in a caravan. Much imagination is now needed to envisage the scene in Rolt's time – he knew Banbury well between 1939 and 1950.

It is necessary to walk through the Castle Quay Shopping Centre until the back entrance opens out onto a paved area, with older buildings and a lift bridge ahead, and levels much changed on either side. This is where Rolt drove in and first saw what he then called the 'Banbury Boatyard'.

On the way to the lift bridge, Tooley's Yard can be seen on the left.[77] The smithy, dry dock and yard were scheduled as an ancient monument in 1975, which protected them from redevelopment.[78] The yard continued working until 1995, after which it was partly rebuilt and surrounded by the first part of the Castle Quay development. This could prompt a significant debate over conservation.

The Oxford Canal and the bridge remain, but the surroundings have changed greatly. The yard occupied the same site, but with buildings that looked makeshift, while the area north and into the distance was an open one, with a branch into Castle Wharf. Factory Street itself formerly continued over the lift bridge to serve further buildings, while the offside area to the south included Banbury Wharf. Most of this is unrecognisable: the Castle Quay development is of a scale and mass that would suggest that it had replaced tall warehouses, but there were none on either side of the canal here.

The revived yard, with the dry dock re-roofed and clad in modern materials, can be seen from above: a covered walkway which is part of Banbury Museum enables the top of the yard buildings to be seen. This view would not be available from the towpath opposite or from the front entrance to the yard. This now seems to sit in what is a very cramped site.

One view is that its character, as a somewhat messy boatyard in a back street, as many canal boatyards were, has been taken away. Another is that because the smithy and dry dock have been kept and are both still in use, along with a boat-repair business that is still there, it is a development of the previous character. The cramped site – almost as if it was an afterthought to retain it – with very large buildings behind, makes it incongruous, but there is a case for celebrating that incongruity, rather like archaeological remains in the middle of a large office development, showing that there was something there before. This would be the case for the canal in Banbury: Banbury Castle had been destroyed in the Civil War, but the Castle Wharf basin and the area around Factory Street, all now under Castle Quay, included much building over the castle site.

North along the towpath, in what is a very enclosed area popular with visiting boaters, is a large modern bridge that carries Cherwell Drive, which provides links to major roads leading to the M40 that were not envisaged in Rolt's Banbury. Since 1999, this bridge has been called Tom Rolt Bridge. His views of such a bridge may be imagined but at least, along with a blue plaque at Tooley's Yard, it has preserved some memory of him in a canal-side landscape that has been so massively changed since the 1940s. It is, after all, where the classic journey in *Narrow Boat* began and where contact with the Tooleys inspired interest in what had been a very private, hidden world of boatbuilding and repair.

There is much more to see in canal-side Banbury after Rolt had passed through, including the lock that was badly damaged in one of only two air raids in September 1940. Returning to Market Place through Castle Quay shopping centre, it is worth heading for High Street, and specifically No. 84, which was Bill Trinder's radio shop. At the time of writing, this is a barber's shop called The Men's Room. A plaque on the side of this building, placed in March 2015, acknowledges both Trinder and Rolt's role in preserving the Talyllyn – a pedant might stress that the two used to meet in a cafe nearby, where the Talyllyn revival was discussed, but the point is well made. Trinder would be a crucial figure, his Liberal connections enabling him to convince the patriarchal Liberal MP, Sir Henry Haydn Jones, who owned the Talyllyn Company, that there was a serious possibility of reviving the railway. One consequence was, presumably, that Jones made his wish to retain the railway clear, so that this could be considered posthumously. It is likely that without Trinder and Rolt the line would have been sold for scrap.

Banbury, then, can be said to be the nearest to a birthplace of two movements: towards the revival of British inland waterways and the preservation of steam railways. A market/railway town when Rolt stayed there, one about which he expressed significant doubts, it would later expand greatly as a London overspill town. Within a short walking distance are two key sites – one quite prominent, one very much hidden and obscure. Ironically, the latter is no longer so obscure, due to the large shopping development that has changed the townscape so much.

Part of the interior of Tooley's Yard, with the covered dry dock to the right. Note the incongruity of the older shed bridged by a modern structure which is part of Banbury Museum.

Looking south towards the lift bridge that carried Factory Street, with Tooley's Yard and Castle Quay on the right. The historic narrow boat on the offside of the canal was built in 1940 (not at Tooley's) and was used for the carrying of coal to Banbury into the late 1950s. Tooley's Boatyard Trust is hoping to restore this vessel.

The entrances to Tooley's Yard, with the museum structure and Castle Quay in the background. Note the blue plaque, which commemorates the start of Tom Rolt's voyage here.

Tom Rolt Bridge in Banbury carries Cherwell Drive and links to the A4260 and A422 roads, which lead to the M40. It seems doubtful that Rolt would have approved of these, but at least the name of the bridge provides a reminder of his involvement.

This plaque by the entrance to the side of Tooley's Yard commemorates the start of Rolt's journey in *Narrow Boat*.

A view of the converted narrow boat *Poo-Koo* moored opposite Tooley's Yard at some date between 1957 and 1962. The contrast between this view from the towpath and the present scene is manifest. (From the collection of C.N. Hadlow and supplied by The Waterways Archive)

*Opposite:* This view of Factory Street, west of Liftbridge 164 over the Oxford Canal, dates from 1960; it was not so decrepit in Rolt's time. On the right, just beyond the lamppost, is the entrance to Tooley's Yard. Banbury Wharf buildings were to the left. (From the collection of C.N. Hadlow and supplied by The Waterways Archive)

# 8

# RAILWAY REVIVAL, CONSERVATION AND VOLUNTARISM

Tom Rolt's explorations of narrow-gauge railways in Ireland in 1946 followed his interest in smaller railways and the workings of larger ones. This was well established by the early 1930s, although his attention had declined during the Phoenix garage period in the mid-1930s.[1] This partly revived during wartime, when in 1943 he eschewed his annual holiday at Llanthony and went to stop near the Talyllyn Lake, under Cader Idris.

The Talyllyn Railway was not his chosen destination – it was only after 'a couple of days' that he realised that its passenger terminus at Abergynolwyn was just 3 miles away.[2] While Tom had visited several narrow-gauge railways before 1939 – two of which were now moribund – he had never seen the Talyllyn.

Seeking to take the Wednesday passenger train back to Abergynolwyn, he and Angela took the bus to Tywyn. He was unimpressed by this seaside resort, although he saw it in driving rain. 'The small towns of real Welsh Wales seem bleak and dreary,' he asserted, while he condemned 'the funereal hues of the dour nineteenth-century houses'. He and Angela (who is not mentioned in *Railway Adventure* – what she thought of it seems to have gone unrecorded) found that trains had been cancelled and decided to walk up the 7 miles of the line; as he put it later, they thus learned more about the railway than if they had travelled over it.[3]

In the same year, James Boyd, a textile manufacturer and rail enthusiast from Manchester, proved to be the sole passenger going up one Friday morning. He described the decrepit 1866 locomotive *Dolgoch* drawing a single passenger coach with three empty trucks behind it. He was allowed to drive the train beyond Abergynolwyn, along to the foot of the quarry funicular, and recorded that only eight men, loading around four wagons a week, now worked at the quarries; pre-war, he was informed, there had been 180 men and three trains of ten wagons daily.[4] The quarries would suffer a disastrous collapse in 1946, after which there was a summer passenger service on three weekdays only, which mostly served holidaymakers.

The railway had been seriously run down, relying on the makeshift maintenance of track and locomotives since Sir Henry Haydn Jones, the local MP, had acquired it along with the quarry leases. The Talyllyn would seem to present an odd choice of railway for one who would seek to preserve 'the pride and efficiency of the old pre-grouping companies, the spit-and-polish of their locomotives and rolling stock which spoke so eloquently of that tremendous *esprit de corps* that existed throughout the railway service'.[5] However, the main grounds for its choice for preservation lay in the omission of the Talyllyn from both grouping and nationalisation, and thus the practical possibility of a transfer of private ownership and control. What might have been chosen, had there been a free hand to transfer control of any railway line, must be left to speculation.

It is noteworthy that Rolt was already drawing far into the past – it was aspects of the railways of the pre-grouping era, before 1921–22, that he wished to conserve or restore. Quite how this would be achieved with the aid of volunteers, many of whom would be middle-class enthusiasts, was another matter.

## Pioneers and Preservation

Advocates of support for Britain's canals up to the 1930s had favoured their development from existing lines to provide better transport facilities, as well as constructing new lines and facilities. There were few examples of the conscious preservation of any waterways, structures or artefacts for historic interest or significance, bar perhaps the conservation of historical records for archives. This was not, however, the case with railways. There were those who advocated the modernisation of lines, methods and rolling stock and the construction of new facilities, but there were also those who sought preservation with various motives.

The earliest instance of railway-related preservation seems to have been the original locomotive *Rocket* – this had been intended to appear at the 1851 Great Exhibition but was donated to the Museum of Patents, which opened in 1857.⁶ Stephenson's *Locomotion* was mounted on a pedestal at Darlington at much the same time.

Any decisions to preserve were capricious and arbitrary – many other old locomotives and items of historic interest were simply cut up and destroyed. Some individuals would attempt to preserve relics: two of the North Eastern Railways' senior staff were especially keen, but by contrast, Stanier, the Chief Mechanical Engineer for the London Midland & Scottish at Swindon, would scrap historic locomotives at Derby from 1932. Publicity provided one motive for commercial preservation, like the Stockton & Darlington's 50th anniversary and then its centenary in 1925; the latter gave rise to the Railway Museum that opened in York in 1928.

In some instances, the age of rolling stock and ramshackle lines, methods and operations made some lines, especially narrow-gauge ones, almost like working museums. Their only viable future could lie, as with the Talyllyn, with their conservation as historic lines that accessed scenery and historic environments for tourists.

Enthusiasm for railways grew in diverse manner, from interest in engines to interest in historic lines, railway history and exploratory travel, encouraged by magazines like the *Railway Magazine* from 1897. The Railway Club, founded in 1899, catered more for those who were professionally involved, but some of those were also rail enthusiasts. The Stephenson Locomotive Society, formed in 1909, developed interests in historical subjects, which could lead to preservation, as with its financing of the restoration of the locomotive *Gladstone* in 1927 for the Museum at York.⁷ The Railway Correspondence & Travel Society was founded in Cheltenham in 1928; Rolt was an early member. These provided talks and ran special outings trains for members and others.

Enthusiasm did not yet extend to the preservation of lines or services, but some element of these was perhaps represented by the conversion of the Ravenglass & Eskdale line (R&E). A narrow-gauge railway carrying iron ore, and later, passengers, the latter kept the line viable until closure in 1913. Revival was carried out from 1915 but using miniature locomotives on a re-laid track at a gauge of 15in. This was 'rescued' by a lease that seems to have been unlawful, granted by a creditor who had no power or right to do so.⁸

The rescue was carried out by Narrow Gauge Railways Ltd (NGRL), a company established in 1911. Until the 1914 war, it supplied and ran exhibition lines abroad and miniature seaside lines at Southport and

Rhyl. NGRL would also be involved in 1916 in the conversion of the Fairbourne Railway, north of Tywyn on the Cambrian Coast, from narrow gauge to miniature. The rolling stock mainly came from moribund enterprises, but two locomotives were bought from the Eaton Hall estate in 1916 – including *Katie*, which may have been that which inspired the infant Tom.[9] The NGRL owners were described as 'professional amateurs, enthusiastic eccentrics, or vice versa'. Financial problems led the R&E to be bought by a Liverpool shipowner and run from that city from 1925.[10] The line was very rundown when Rolt travelled over it in 1947.

The revival of the R&E, on shaky financial ground, was not really an act of deliberate historic conservation of line, structures or rolling stock, although it did preserve the route and passenger services; indeed, the passenger line was extended. Although the first NGRL directors were rail enthusiasts, there was no support society and no use of volunteers.

Holmes has stressed that there were several narrow-gauge lines in existence in 1915–17, so that any loss of the R&E would not have concerned many rail enthusiasts at the time. However, by 1948 the main surviving narrow-gauge steam passenger line was in the Vale of Rheidol, running from Aberystwyth, and this was nationalised. Holmes concluded, 'It was the near-unique survival of the Victorian Talyllyn complete with its original 1860s equipment which excited popular interest.'[11]

The revival of many small-scale lines would be negated, in many cases, by the need for their preservation to be authorised by their owner, which would generally require a transfer of ownership. The legal framework for most railway lines made this difficult, as was to be experienced by those who sought to revive two narrow-gauge railways in north-west Wales, the Welsh Highland and the Ffestiniog. Ownership was enmeshed in a maze of shareholdings, loans and creditors. Nationalisation presented an unknown future, but the priority would be modernisation with, as Rolt pointed out in 1953, the closure of branch lines in greater numbers than under private ownership.

The Talyllyn had the advantage that it had a single dominant ownership, whereby rights could be transferred. The body which took it over was unique at the time.

The idea for the Talyllyn Railway Preservation Society seems to have come from Owen Prosser, who would also be interested in campaigning for the better use and development of existing and new railways.[12] The idea of reviving and preserving a declining railway line might later appear uncontroversial, but detailed politics need to be considered – what was being conserved, how, by whom and why?

## Talyllyn Railway and Survival

An analysis of the politics of revival could be first approached by considering the kind of ownership: what caused the Talyllyn Railway to be conserved, in engineering and human terms?

The railway had been built for the McConnel family of textile manufacturers, who diversified into quarry exploitation (which the railway served) when manufacturing became less profitable. The venture was overcapitalised and these distant owners were clearly only interested in its capital value, judging it and the quarry against various other assets that they owned. Its sale was a continuous possibility from early on, when new services to Abergynolwyn were announced in 1900.[13] The line was operating at a loss in 1910, when the McConnels, who had already closed the quarries, sought to sell or indeed close it. This prompted the first move for its revival or retention.

The local council convened a meeting of local ratepayers on 12 October 1910 in Tywyn. While it had no powers to acquire or run either, it was concerned about the impending closure of the quarry and line and its effects on tourist income. 'For the last 40 years it had conveyed thousands of visitors,' it was asserted, and it was felt that the visitors that were attracted were beneficial, 'as the residents were dependent on visitors who made frequent excursions by this railway'. Dolgoch, with its falls, was singled out as a major destination.

The income for the last five years from the slate traffic had averaged £650, making a profit, but the railway accounts for the year ending in March 1910 showed a deficit of £484 10s 1d.[14] As closure on 15 November was proposed, a council committee was formed to negotiate with McConnel.

The purchaser and sole bidder proved to be the solicitor Henry Haydn Jones, a local property owner, councillor and newly elected MP for Merioneth, who was interested in maintaining a local venture but with minimum expenditure and capital investment.[15] One motivation may have been the impact of the closure of the railway and quarries on the values of property that he owned. Jones now owned a majority of the shares in the statutory Talyllyn Railway Company, and after 1941 held a series of short leases on the quarries until these ceased operation. He would be well described as autocratic, opposing union negotiators and their involvement, but benevolent in his wish to keep the railway running.

By 8 February 1912, it was reported that the quarries had reopened and the line was running. The surveyor to the council in Tywyn hoped that 'the prosperity of fourteen to fifteen years ago could be revived'.[16] That Jones focused more on the passenger and tourist service is indicated by the new passenger use of Wharf Station, which had previously been confined to the loading of slate and other freight.[17]

Jones had determined to keep the line open in his lifetime, even if this needed personal subsidy. As the Talyllyn Company was not due for nationalisation (it was later considered, but rejected), unlike most railways, one possibility was that a group of rich men, including former railway directors, might join together to acquire it.[18] This was a solution put forward by Bill Trinder, who would become the first chair of the Talyllyn Railway Preservation Society. Rolt had got to know Trinder since 1939, when he first visited his radio shop in Banbury, through which a friendship developed.[19]

## Enthusiast Moves to Revival

Rolt's regret at railway nationalisation, shared with many Conservatives (and some Liberals, like Trinder), provided one spur. Aware that most railways in Britain would be nationalised, with the apparent loss of individuality that this was perceived to bring, he later reflected, 'What a fine thing it would be if at least one independent railway could survive to perpetuate, if only upon a small scale, the pride and the glory of the old companies [...] But then I thought of that seven miles of worn out track'.[20]

Rolt's 'old companies' were those pre-grouping, and indeed pre-General Strike. Raymond Williams' father, a signal worker, was among many victimised by the Great Western Railway Company, somewhat dampening 'the pride and the glory' for those employed. It was in the pre-grouping era in the mid-1870s that annual deaths at work of railway employees peaked at 700.[21] Curiously, the Great Western Railway had tried to close the Kennet & Avon Canal, and had neglected the Stratford Canal, both of which it owned. However, Rolt was clearly keen to save one railway to demonstrate and celebrate links with a favoured era, although others might view that past differently. The Talyllyn presented a rare candidate.

In 1947, Trinder suggested that Sir George Shouster, a former National Liberal MP for Walsall who had lost his seat in 1945, might be willing to gather a group of wealthy fellow enthusiasts to bid for the railway. Trinder knew Shouster, who was a director of the Southern Railway Company, through his involvement with Liberal Party politics. Rolt's reported response was that 'he would rather have hundreds of railway enthusiasts making a small contribution of a few pound than tapping half a dozen wealthy men to take it over'.[22]

It is unclear what view Rolt took of Owen Prosser, who was interested in acquiring the railway, presumably to pass it to some sort of trust. Soon after Sir Haydn died on 2 July, he made enquiries of Edward Thomas, the manager of the line. Prosser had been influenced by a 1941 letter in *Modern Tramway* magazine by an enthusiast, Arthur Rimmer, who had suggested that existing railway clubs and enthusiasts might acquire and revive the Welsh Highland Railway, then derelict and about to be dismantled. This proposal did not necessarily involve the recruitment of voluntary labour.

Prosser was then a rare example of an enthusiast for historic and modern tramways, waterways (he attended the Market Harborough Rally in 1950 by canoe) and both modern and historic railways. The latter focused mainly on the retention and revival of branch lines, many of which were soon to be closed under nationalisation. He would suggest the founding of a Talyllyn Railway Preservation Society in a letter to which Rolt replied on 22 September 1950.[23] Curiously, Rolt made no mention of Prosser in either *Railway Adventure* or *Landscape with Figures*.

Prosser's proposal was overtaken by the public meeting on 11 October 1950 in Birmingham, a suitable location with the meeting attended by members of the Stephenson Locomotive Society and Birmingham Model Railway Club. This led to the formation of the society at its first committee meeting there on 23 October. At that point, the proposal was to investigate possibilities, should Haydn Jones' widow agree that the new society could have control of the railway. This would prove to be the case, with a formal agreement in February 1951.

## Voluntary Organisation and Revival

The new society would provide an unusual structure to control the railway and run services, one that many later revivals would not follow.

In effect, the members of the Talyllyn Railway Preservation Society (TRPS) elected the committee that actually ran the railway, appointed and employed the few paid staff and supervised the many volunteers. This has been described as a 'workers co-operative', but this is inaccurate – the paid staff did not own the company or have any say. It might be closer to a producer co-operative model, whereby leading members run the company in the (manifest or purported) interests of the members.

During Rolt's time as manager, he and Sonia, now his partner, along with the former staff from the Jones era, were the main employees, while the committee comprised mainly businessmen from the West Midlands – from local captains of industry to builders like Patrick Whitehouse and accountants who serviced these owners, like Patrick Garland. Only one, John Wilkins, had experience in running a railway, and that a miniature tourist line. As Rolt noted, 'None of us had had practical experience of running any railway larger than a miniature or a model.' 'For a small band of amateur railwaymen to undertake the operation of even the smoothest running of railways would have been venturesome enough,' he stressed, but for one as eccentric as the Talyllyn it made the enterprise an 'adventure' indeed.[24]

These figures were experienced in management, had practical resources and access to more, while Wilkins (a manufacturer of washing machines, inter alia) had a significant ownership in the nearby Fairbourne Railway. This had been revived in 1946–47 without voluntary funding or effort.

Rolt was to chafe at the power relations involved, and their practical consequences, which he regarded as interference with his efforts as manager on site. These formed a major factor in his decision not to continue into a third season after 1952. The interests involved were more like the running of a members' club, pooling resources with a voluntary but powerful management.

Dunstone has characterised this movement as 'enter private enterprise', although it might be better seen as 'voluntary enterprise'; there was no capital invested as such.[25] Certainly, there was a contrast with the national state ownership brought about by the Transport Act 1947, with the unification of ownership, associated bureaucracy and rationalisation and withdrawal of services. Like the Corris Railway, had the Talyllyn been nationalised, it would almost certainly have been closed early on.

The vicissitudes involved made it fortuitous that so many enthusiasts proved willing to volunteer to assist, given the practical difficulties that would have been involved in organising paid staff to improvise solutions for an appalling track and rolling stock.

Precisely what was to be conserved? Clearly, the kind of commercial service that the McConnel firm had promoted, mostly for the carriage of slate but later with local passengers and tourists, could not be revived, although much of that line, notably engineering and locomotives, still remained. In their time, the passenger line ran only to and from Pendre, at the north end of Tywyn, although it was possible to board at the 'slate wharf' in later years.

Partly inspired by a 1947 visit to the R&E, surprisingly, Tom felt at first that to relay the Talyllyn, with its appalling track, to miniature gauge was one way of conserving it. This would keep the short section between Wharf and Pendre Stations to accommodate existing locomotives and rolling stock.[26] This would not have conserved much of the historic character of the railway or provided any continuity with the service running in the 1940s, yet it would have conserved the route, some structures like bridges and viaducts and the ability to travel up the valley. Others advised him that this would have deterred much public interest and commitment; the appeal would lie with an historic line.

One feature of this is that it could have contributed to that intangible heritage of 'pride and efficiency' of which Rolt approved, even if it was running small-scale locomotives over miniature lines. At least an efficient service could be sought more easily than with a railway with heavy engineering, and this might contribute to a sense of pride, albeit that it might be felt (and transmitted) by middle-class volunteers giving up part of their holidays.

The service which was to be revived and modernised, to conserve the engineering and some parts of its operation, became largely that from the Haydn Jones era. This transfer involved, not the continuation of an existing service over a ramshackle line but its extensive restoration, revival and transformation into a line devoted largely to tourism. Its main function, conveying slate, had ended and, as Rolt soon discovered, any quarry revival was unlikely. Local passenger carrying had dwindled since the introduction of bus services, with only occasional parcels and supplies, so the only commercial future lay with the development of tourism. Rolt did wonder later whether he might have been better to leave the line to slumber: 'This lost railway had a certain magical quality about it which makes me wonder sometimes whether we did right to disturb it.'[27]

These misgivings reflected the problems of much conservation of 'heritage' – the neglect that had saved this feature from modernisation and left elements relatively free for preservation, and kept it hidden, meant that only promotion and publicity, notably for tourism, could lead to its revival and retention. Many qualities would be lost, as Rolt found with historic canals.

## Voluntary Management, Labour and Finance

The politics to which much of the Talyllyn conservation corresponded rested on an apparent assumption that voluntary management, assisted by voluntary labour, mostly by railway enthusiasts, could achieve what state ownership could not: the survival of a fairly viable private passenger railway. This would not have access to private capital as such, although it is doubtful that any private shareholders would have invested. It did mean voluntary contributions and some access to cheap, enthusiast subsidised repair and reconstruction facilities. This was less in the vein of conventional fundraising than in the provision of private works for minimal or no cost to supportive English Midlands industrialists and other contacts. These included Wilkins, who supplied materials and labour from the Fairbourne and left it for the Talyllyn committee to pay as funds became available. There were also some charitable donations – not just the TRPS membership fees, but ones like a donation of £750 from the Newcomen Society in America.

Voluntary labour was also soon involved and working parties began almost as soon as control was handed over in March 1951. These seem to have partly reflected traditions of voluntary service by parts of the middle class, turning enthusiasm to practical effect, much of it from distant locations. It is not clear precisely how these were organised with the specific skills involved. Owen Prosser, who claimed to be at the first working party, recorded track work by men wearing collars, ties and even suits. Prosser had already worked on the Lower Avon Navigation, but that included vegetation clearance, patching brickwork and timberwork to lock gates, not the preparation of a functioning public railway line. It seems that the work was set out by the small permanent staff, but how it was supervised and how safety standards were assured for a public railway that carried passengers is uncertain.

The formation of a voluntary body to run a railway was a highly unusual development in 1950. Some at the Birmingham founding meeting doubted that it would come to much.[28] A distinction needs to be made between a voluntary organisation and one that makes use of voluntary labour. A voluntary organisation is founded by its members, does not have to exist and does not have to have charitable status, despite some fiscal and legal advantages. In principle, a statutory body could make use of volunteers – parts of the NHS and local social services have made such use, often dealing with voluntary bodies that organise and 'employ' volunteers. The use of voluntary labour or voluntary funds to secure railway or waterways preservation seems to have emerged only in the 1940s with Rimmer's letter of 1941. Funding – raising money through regular subscriptions, donations or legacies – provides another voluntary element, along with sometimes charitable funding and specific government or charitable grants for particular purposes.

In the late 1940s, a major change to the practice of voluntarism was the nationalisation of hospitals in Britain, which removed a substantial part of voluntary fundraising and management. One view (probably close to Rolt's) was that this represented the surrender of the voluntary principle to the bureaucratic state; another, that this freed up voluntary commitment to pass to other interests.

Rolt recorded an element of defiance against the rule-bound nature of union-backed work – 'that's not my job' was never heard among volunteers, he wrote proudly.[29] *Railway Adventure* does not mention military involvement, using Territorial Army units, which would follow only after Rolt had ceased to be manager.[30] There was an element of enthusiasm and tolerance for improvisation in many passengers, who were largely visiting tourists. After the vicissitudes of wartime and post-war shortages and enthusiasm and goodwill for the enterprise, these must have seemed acceptable. *Railway Adventure* records numerous breakdowns and improvisations, which would not be tolerated later, and the need for investment in the restoration of the engines and track was clear.

One early consequence of nationalisation was the closure of the nearby Corris Railway following flood damage. As this line had the same gauge as the Talyllyn, the opportunity arose to acquire one or both locomotives. Rolt recorded the visit to Swindon, to the former Great Western works, at which Trinder's attempts to strike a deal with British Railways representatives seemed to founder against a new bureaucracy.

Although Rolt fulminated against the loss of 'the succession of famous men who had done so much to make the Western Railway Great', he did not consider that a previous autocratic Chief Mechanical Engineer, Churchward, had ordered the destruction of a number of historic locomotives at Swindon in the early twentieth century.[31] In the event, judicious adjustment of the book values of these old locomotives meant that the Talyllyn could acquire both.

## Dilemmas, Tourism and Conservation

Rolt expressed unease at the deep reliance of the railway upon tourism and was scornful about some of the visitors' attitudes involved. For instance, he recorded vandalism by pupils from a girls' school, perhaps of a similar nature to the school he had attended.[32]

Local bus competition had reduced passenger carrying, never significant on the Talyllyn, to a summer tourist service. While Rolt managed the Talyllyn, he favoured the very small numbers of 'genuine' passengers to and from Tywyn and occasional parcels and small goods up the line. At busy times, he carried the few remaining local passengers in the brake van, and 'it was as though our brake van had become the last stubbornly held strong point for a local underground resistance movement forced to take desperate defensive measures'.[33]

His language revealed feelings of despair: what Rolt valued was so fragile. He realised that tourists, with 'the loud, harsh accents of London, Lancashire, Birmingham or the Black Country',[34] would dominate, and recorded wearily the over-eagerness of rail enthusiasts and the facetious remarks of holidaymakers.

The railway's dissonant heritage is indicated by the Welsh workers who were employed on its locomotives before the voluntary takeover. Far from ensuring continuity of operations, they walked out and subsequently tried to discourage volunteers. Rolt later suggested that they left deliberately, so as to sabotage the voluntary operation.[35] He did not consider the poor wages, long hours, primitive methods and working conditions on the Talyllyn, or what the workforce saw as interference from middle-class volunteers – people who could afford holidays ordering around workers whose wages might not support holidays.

Rolt saw the railway as more than an interesting partial survival from the 1860s accessing a picturesque landscape:

Wales is the ages old association of people with landscape in an ecological partnership intimately interlinked. To the landscape the people have given the sheep, the black cattle and the small hill farms; the dry-stone boundary walls and mountain quarries; the little market towns and sea-ports. To the people the mountains have given a way of life and, until the twentieth century invasion, they have nursed and protected the language and the traditions of a people which are the expression of that way of life.[36]

It is hard to agree that this organic vision was based on an 'ages old association', given the local economic disruption that the quarry and railway had created in the mid-nineteenth century. From the outset, slate was largely exported by railway and sea, serving the needs of distant urban places over local needs.

Despite his admiration for craftsmanship, Rolt failed to commend the craftsmen of the quarry industry. They had dominated north Wales' slate-quarrying operations, exercising detailed controls over work processes in quarries like the Penrhyn, where the long strike of 1900–03 resulted as Lord Penrhyn sought greater control.[37] Praise for the independence and influence of the *gwerin* (folk) did not extend to their political demands or the revival of their work processes.[38]

He was presumably impressed by the maintenance carried out on locomotives, the earliest from 1865, which had been kept going by the small staff at Pendre. There was praise for the improvisations produced by the many enthusiastic volunteers, but little seems to have been recorded about the way in which they acquired and developed their skills. Much was made – as has been stressed elsewhere – of the varied backgrounds, mostly from the middle class for whom manual work was unusual, such as 'undergraduates, shopkeepers, clergymen, engineers, railwaymen and schoolmasters'.[39] These were hardly representative of the British public, but there was praise for those with railway experience, some of it from working for British Railways, who gave up their spare time to help on the railway.

Rolt, unusually, emphasised the problems of white-collar work and administration, perhaps because much of this was left to him, as manager, or his partner Sonia. There is no discussion of administration in his canal work, such as *The Inland Waterways of England*. He commented, 'For most people the booking clerk is no more than a disembodied hand which scoops up their money,' but pointed out the difficulties for volunteer guards who, like him sometimes, issued tickets on the train itself.[40] He emphasised:

> As this was a department of railway operation of which I had had no previous experience or knowledge whatsoever, and as we did not number amongst our members any railway booking clerk to instruct me in the art, I had to proceed on the principle of trial and error until the procedure was mastered and became a matter of routine.[41]

It is instructive of the enthusiastic amateurism that attended the Talyllyn that someone with no experience of railway management or clerical work (beyond extensive correspondence in his role as IWA honorary secretary) should be appointed to run a railway with engineering and administrative responsibilities. It was fortunate that Tom was able to adapt quickly, and that his successors had relevant experience.

## Some Conclusions

In 1992, the late John Snell, who worked voluntarily at the Talyllyn between public school and university in 1951, summed up several ideological factors that underlay the Talyllyn revival:

> I certainly had a typical teenager's cause myself: if we could hold the fort at such a threatened outpost as the Talyllyn Railway, surely no one would have the arrogance to dream of closing such useful branches as those to places like Tenterden, or Bellingham (North Tyne), let alone important main lines like the Somerset & Dorset. Tom's cause was, it turns out, more realistic and achievable: to reverse the fashion to centralise and nationalise; to restore the tradition of self-supporting local concerns in which people could take a legitimate pride in achievement. In fact, to work towards some of the ideals that later grouped under the banner 'Small is Beautiful', but more than that: to enlarge the duty, and the joy, of living.[42]

These were extravagant claims, but there is something of Rolt's feeling behind them. There is much to criticise: all the railways named by Snell did close – the Somerset & Dorset was the last, in 1966. It is hard to see the Talyllyn as self-supporting locally, despite the free labour of so many supporters from distant areas like the Midlands. Rolt himself realised that the line was losing money while he was manager.[43]

Snell's idealistic emphasis on the 'joy, of living' perhaps reflects the views of some volunteers, who were taking great satisfaction in forms of work over which they had much control – notably, the ability to work with fellow enthusiasts and to work on something that mattered to the volunteer. This might, indeed, produce pride and *esprit de corps*. However, this does attach ideological baggage to activities which might seem to attract a broad consensus – harnessing oneself to a kind of opposition to nationalised transport without any critique of the kind of employment relations, tasks and pressures that attend most private sector enterprises in neoliberal capitalism. Snell seems to have evoked views that opposed trade unions and were against any sort of public ownership. Rolt might well concur with this, but some who visit and admire heritage railways might not subscribe to such ideologies.

While much that happened to Britain's railways took place in Tom's later and final years, there would be a long-term decline in the route miles of the nationalised system and the loss of many freight traffic and passenger services. Some of this drew opposition – in the case of Owen Prosser, from the outset, when branch line closures were first proposed in the late 1940s.

One response to this was to retreat into small-scale revivals, sometimes including the restoration of lines that had been dismantled. Many small-scale schemes did follow the Talyllyn's example, so that fifty years later, there were 400 miles of revived heritage lines. Can these small-scale operations, none of them providing a mainstream everyday passenger service, be seen as conserving part of the everyday railway?

Another approach would have been closer to that of Robert Aickman with waterways – to proclaim that the entire rail system must be retained and the loss of any part was a betrayal of the whole cause. A smaller version would be to oppose the threatened closure of the Cambrian Coast Line in the early 1970s. Here was a line that still had freight traffic and was in regular passenger use. It was making a loss but contributed to local tourism and connectivity. The line did stay open after a lengthy campaign, but how far this included TRPS members or was inspired by the Talyllyn enthusiasm is unclear.

While the Talyllyn and similar lines provide a pleasant holiday experience, they rarely provide a realistic local service. To protect historic structures, staff stations and maintain historic rolling stock as a living museum, a railway may need to be run by enthusiasts. However, the Talyllyn experience is ambivalent if it is regarded as recreating past journeys. The Wharf Station has been largely rebuilt since Rolt's time as manager when it had no platform. Abergynolwyn Station has also been reshaped, with the extension to Nant Gwernol that was initiated by Rolt in 1970 but completed after his death. The rolling stock is historic, but only the 1860s locomotives 'belonged' to the line; the Corris ones did not, while the locomotive *Tom Rolt* is based on one from a former Irish peat railway. If there really was an intention to revive some sort of 'pride and efficiency', it would need to be from paid regular staff, not enthusiastic volunteers.

None of this is to diminish the attractiveness of the Talyllyn or the collective efforts and commitments over seventy years, but it is the claims made that are inconsistent. Tom may have been right in wondering whether the line should have been left to slumber.

As with the IWA, a similar question arises. What if Tom Rolt, considering his broken marriage, falling bank balance and aversion to taking orders, had declined to take on the management of the Talyllyn in 1951 and 1952? It is difficult to imagine that anyone else would have come forward who was willing to try to keep going what was, by all descriptions and especially his, a neglected, makeshift shambles of a public railway. It would have been inconceivable that anyone could have designed a job description that included, for instance, taking one's own motor car in case passengers needed to be rescued from an ailing train or coaxing a derailed locomotive back onto the rails.

Most enthusiasts had little technical or administrative knowledge to back the running of a railway service, and those who did would often be in full-time employment. Rolt's successor was a retired railway manager, but by that time enough problems had been overcome to form some sort of routine management solutions. Once again, a great deal is owed to Rolt and, had the Talyllyn failed, as well it might, many later schemes might never have come about.[44]

On 15 October 2011, the 1866 locomotive *Dolgoch* featured the headboard 'The Rolt Explorer'. It is seen here at Rhydyronen Station, the first up from Pendre.

At Wharf Station in 1951 with Rolt's Alvis on the right. Note the lack of terminal buildings. (Ian L. Wright, courtesy of Stephen Rowson)

*Opposite: Dolgoch* is seen here at Brynglas passing loop, 15 October 2011.

# 9

# AFTERWORD

This is a book of essays rather than any comprehensive study, and conclusions must be provisional. The selective nature of essays means that there is much scope for others to produce commentaries based on research into matters that interested Tom Rolt.

I have only covered some aspects of Tom Rolt's life and work up to the early 1950s. There are several subjects that could have been covered during this period, and indeed, many about his later life and involvements. I have been concerned here with the Rolt who was influential in securing the revival of canals and the Talyllyn Railway, not the later full-time writer and engineering biographer, involved in museums and industrial archaeology. I will have to leave it to others to confirm or deny my feeling that the later Tom had slightly different perspectives and priorities; certainly, he had a different impact.

I am drawn to the oft-made conclusion that further studies are both possible and desirable. Tom Rolt's name still has the capacity to inspire, but fifty years after his death, what more can be learned? One question is about material that can be discovered in archives. These essays have been written without access to the extensive materials at Ironbridge, which were fully catalogued in the summer of 2023; these might reveal details over which there has been, for the present, speculation.

Although it relates to a later period, when Rolt was a member of the Inland Waterways Redevelopment Advisory Committee at the end of the 1950s, my own voluntary cataloguing work at Ellesmere Port turned up a set of letters about the Dudley Tunnel which had not hitherto been seen, to my knowledge. There may be materials elsewhere that have not yet been revealed. Whether these might modify broad conclusions rather than add details remains to be seen.

My essays may well have shown that there is much with which some would disagree markedly. However, for me, Rolt opens windows on many issues that are worthy of investigation. These also open on to a series of pasts – providing portraits of waterways, industries, craft and of course, past landscapes. If his perceptions, deductions and conclusions appear sometimes to be questionable, they can still prompt discussion and dialogue. And it is certainly worth reflecting on places that mattered to him, and how his historical portraits show these have changed.

Rolt is one of several figures born between 1909 and 1914 who have had an impact on later understandings of transport history and industrial archaeology, in both practice and writing. I would repeat my suggestion, made twenty-five years ago in *Charles Hadfield: Canal Man and More*, that if the detailed history of movements towards transport history and industrial archaeology in Britain is ever written, the varied and interlocking parts played by these figures – Charles Hadfield, Tom Rolt, Thomas Willan, Jack Simmons, Alec Skempton and Michael Rix – need to be considered. Other names may well be added for a collective biography, along with an additional scope – their relationships to the conservation of heritage, tangible and intangible, historic environments, nature and resources.

The ability to access those who knew these people may have diminished but the results of research might well be fruitful. A collective biography would need to set such figures in the context of events – itself founded in structural forces that conditioned those events. These essays offer a small contribution to such studies.

It would be easy enough to dismiss Rolt's contemporary writings on environmental themes as fit only for antiquarian studies and far distant from present concerns. However, this would be to miss the prescience which they displayed.

He found himself on the organicist wing of rural policies, but nature and historic conservation practices and principles have turned very much towards the planner-preservationist rather than the organicist. This has applied to the growth of Green politics, which were very much organicist and apocalyptic (and uninfluential) in Rolt's last years but have since moved much closer to mainstream acceptance in British politics. This has been much more about modernisation in a planner-preservationist fashion than about making a return to a small-scale, self-sufficient past, which was in many ways a legacy of the organicism of the 1930s and 1940s that he had encountered, and to which he contributed.

Rolt's analysis of the impact of industrialism and the disruption of organic relationships may lead readers to query just how 'organic' social and economic relations might have been in the past, but the call for more 'organic' relations between societies and environments in the future may have resonances. There may well be less to gain from an attempt to return to a past that featured so much deprivation than to work towards a future in which humans work creatively within natural limits, rather than seek the conquest of nature.

If Rolt was wrong to envisage that a totalitarian society was being brought about by the faltering steps towards social democracy in post-war Britain – which had their impact on canals and railways – he might have been on surer ground if he had been able to consider neoliberal capitalism and its corresponding authoritarian movements and governments. The notion that the reader can represent a rhetorical 'we', who can make choices over the kind of society in which we live is familiar enough in literature, but it was one of Rolt's weaknesses that he was unable to discern or consider the binding nature of social and economic relationships. It is poignant, however, to consider his optimism in wartime that a more harmonious society would emerge after the war and how that was shattered by the atomic bombing of Japan and other wartime revelations.

Many impacts of Rolt's writing and practice were unintentional – for both waterways and the Talyllyn, he helped to inspire and facilitate tourism rather than the protection of what he saw as valuable communities. He attempted to highlight survivals in places and their associated crafts but was unable to save them intact. The turn towards the preservation and revival of crafts as leisure activities and part of artistic production came too late for Rolt to consider.

This relates to the idea of 'heritage', which is much maligned in some discussions. The conservation of any heritage must be selective, and in his time, in Britain, this turned from ancient places to scenically valued landscapes and the remains of the Industrial Revolution, with canals and railways among them. If this is a valid selection of heritage, Rolt was among the early exponents. If there were doubts about some of his proposed methods of conservation – as in the miniaturisation of the Talyllyn – this was at an earlier time when there were fewer established methods in the conservation of railway heritage.

One feature in which he was a pioneer was in intangible heritage – his desire not only to conserve boats, engineering structures and physical landscapes, but also to conserve ways of life, specifically those who lived-in on narrow boats. His writing suggests that he saw landscapes as living, working landscapes in which narrow boats would not only be set, but which would reflect their presence. Pleasure boats could not engender such landscapes; for that matter, neither could train services that solely catered for tourists.

One approach to intangible heritage is to record elements of lives which can be passed on or at least appreciated. Passing on could mean that they become ossified into some false 'tradition' or tourist commodity, but it could involve the transmission of meanings. Today, audio-visual recording is much more accessible, but in Rolt's time, writing represented a means of 'conservation by record'. Even when their meaning can be questioned, such writing enables interpretations of the past that would otherwise disappear.

One of Rolt's main legacies lies in the longstanding value of his writing, which has preserved some of what he perceived and admired, even if much has since changed beyond recognition. Considering these works in depth, and visiting places that he knew, can prove valuable. Without explicit intention, he found places and people and gave away some of his secret interest in them. The people are gone, and places are gone or greatly changed, but his writings provide some guidance towards a passage to the past, and its possible future.

# GLOSSARY

| | |
|---|---|
| IWA | Inland Waterways Association |
| IWAI | Inland Waterways Association of Ireland |
| BTC | British Transport Commission |
| VSCC | Vintage Sports-Car Club |
| LRTL | Light Railway Transport League |
| FCC | Friends of Canterbury Cathedral |
| CPRE | Council for the Preservation of Rural England |
| DIWE | Docks and Inland Waterways Executive |
| SGM | Special General Meeting |
| NGRL | Narrow Gauge Railways Limited |
| R&E | Ravenglass & Eskdale Railway |
| TRPS | Talyllyn Railway Preservation Society |
| SCC | Severn Carrying Company |
| CIÉ | Córas Iompair Éireann |
| SDA | Shannon Development Association |

# BIBLIOGRAPHY

## L.T.C. Rolt

This lists the books by L.T.C. Rolt that are cited in the text. A full bibliography is available via ltcrolt.org.uk/biblio.htm and a list of forty main titles is available at ltcrolt.org.uk/titles.htm.

*Green & Silver* (London: George Allen & Unwin, 1949).
*High Horse Riderless* (London: George Allen & Unwin, 1947).
*Landscape with Canals* (Stroud: Alan Sutton, 1977).
*Landscape with Figures* (Stroud: Alan Sutton, 1992).
*Landscape with Machines* (London: Longman, 1971).
*Lines of Character* (London: Constable, 1952).
*Narrow Boat* (London: Eyre & Spottiswoode, 1944).
*Navigable Waterways* (London: Longman, 1969).
*Railway Adventure* (London: Constable, 1953).
*The Inland Waterways of England* (London: George Allen & Unwin, 1950).
*The Thames from Mouth to Source* (London: Batsford, 1951).
*Thomas Telford* (London: Longman Green, 1958).
*Winterstoke* (London: Constable, 1954).
*Worcestershire* (London: Robert Hale, 1949).

## Other Works

Abelson, Edward (ed.), *A Mirror of England: Anthology of Writings by H.J. Massingham* (Hartland, Devon: Green Books, 1988).
Acton, Thomas, and Gary Mundy (eds), *Romani culture and Gypsy Identity* (Hatfield: University of Hertfordshire Press, 1997).
Aickman, Robert, *The River Runs Uphill* (North Yorkshire: Tartarus Press, 2014).
Armitage, Matthew, *Forging Ahead* (Yarnton: Windlass Publishing, 2018).
Bate, John, *The Chronicles of Pendre Sidings* (Chester: RailRomances, 2002).
Blamires, Harry, *Twentieth-Century English Literature* (London: MacMillan, 1982).
Blythe, Ronald, *A Writer's Day Book* (Nottingham: Trent Editions, 2006).
Bolton, David, *Race Against Time* (London: Methuen, 1990).
Boughey, Joseph, and Charles Hadfield, *Charles Hadfield: Canal Man and More* (Stroud: Sutton Publishing, 1998).

Boughey, Joseph, and Charles Hadfield, *British Canals: The Standard History* (Cheltenham: History Press, 2023).

Bradley, Simon, *The Railways: Nation, Network and People* (London: Profile Books, 2015).

Bryant, Arthur, *The Spirit of Conservatism* (London: Methuen, 1929).

Chappell, W. Reid, *The Shropshire of Mary Webb* (London: Cecil Palmer, 1931).

Clapham, J.H., *A Concise Economic History of Britain: From the earliest times to 1750* (Cambridge: Cambridge University Press, 1949).

Coogan, Tim Pat, *Eamon de Valera: The Man Who was Ireland* (New York: Barnes & Noble, 1993).

Davies, W.J.K., *The Ravenglass & Eskdale Railway* (Newton Abbot: David & Charles, second edition, 1981).

Deane, Seamus, *A Short History of Irish Literature* (London: Hutchinson Education, 1986).

Delany, Ruth, *Ireland's Inland Waterways* (Belfast: Appletree Press, 2004).

Dunstone, Denis, *For the Love of Trains* (London: Ian Allan Ltd, 2007).

Foster, R.F., *Modern Ireland 1600–1972* (London: Allen Lane, 1988).

Gaut, Robert C., *A History of Worcestershire Agriculture and Rural Evolution* (Worcester: Littlebury & Co., 1939).

Gwynne, Bob, *Railway Preservation in Britain* (Oxford: Shire Publications, 2011).

Hadfield, Charles, *Introducing Canals* (London: Ernest Benn, 1954).

Hadfield, Charles, *The Canals of Southern England* (London: Phoenix House, 1955).

Hadfield, Charles, *The Canals of South Wales and the Border* (London: Phoenix House, 1960)

Hadfield, Charles, *The Canals of the West Midlands* (Newton Abbot: David & Charles, 1966).

Hadfield, Charles, *Thomas Telford's Temptation* (Cleobury Mortimer: M. & M. Baldwin, 1993).

Hanson, Harry, *The Canal Boatmen 1760–1914* (Manchester: Manchester University Press, 1975).

Harte, Negley, *The Study of Economic History* (London: Routledge, 2006).

Harvie, Christopher, *A Floating Commonwealth: Politics, Culture and Technology on Britain's Atlantic Coast, 1860–1930* (Oxford: Oxford University Press, 2008).

Higham, David, *The Priests and People of Harvington* (Leominster: Gracewing, 2006).

Holmes, Alan, *Talyllyn Revived* (Tywyn: Talyllyn Railway, 2009).

Humble, Nicola, *The Feminine Middlebrow Novel, 1920s to 1950s: Class, Domesticity and Bohemianism* (Oxford: Oxford University Press, 2001).

Huxley, Aldous, *Brave New World* (London: Chatto & Windus, 1932).

Huxley, Aldous, *Ends and Means* (London: Chatto & Windus, 1937).

Jones, R. Merfyn, *The North Wales Quarrymen, 1874–1922* (Cardiff: University of Wales Press, 1981).

Kiberd, Declan, *Inventing Ireland* (London: Jonathan Cape, 1995).

Klingender, F.D., *Art and the Industrial Revolution* (London: Noel Carrington, 1947).

Llewellyn, Richard, *How Green Was My Valley* (London: Michael Joseph, 1939).

MacDermot, E.T., *History of the Great Western Railway* (London: Great Western Railway, two volumes, 1927, 1931).

Mackersey, Ian, *Tom Rolt and the Cressy Years* (London: M. and M. Baldwin, 1983).
Mais, S.P.B., *This Unknown Island* (London: Putnam, 1932).
Massingham, H.J., *Remembrance* (London: Batsford, 1941).
Massingham, H.J., *The English Countryman* (London: Batsford, 1942).
Massingham, H.J., *Where Man Belongs* (London: Collins, 1946).
Massingham, H.J., *The Curious Traveller* (London: Collins, 1950).
Massingham, H.J., *The Southern Marches* (London: Robert Hale, 1952).
Matthews, Jodie, *The British Industrial Canal* (Cardiff: University of Wales Press, 2023).
Matless, David, *In the Nature of Landscape: Cultural Geography on the Norfolk Broads* (Chichester: Wiley Blackwell, 2014).
Matless, David, *Landscape and Englishness* (London: Reaktion Books, second edition, 2016).
Mullay, A.J., *Railways for the People: The Nationalisation of Britain's Railways in 1948* (York: Pendragon, 2006).
Neff, J.U., *The Rise of the British Coal Industry* (London: Routledge, 1932).
Owens, Victoria, *The Life of L.T.C. Rolt: Where Engineering Met Literature* (Barnsley: Pen & Sword History, 2024).
Pevsner, Nikolaus, and Edward Hubbard, *The Buildings of England: Cheshire* (London: Penguin, 1971).
Plomer, William (ed.), *Kilvert's Diary 1870–1879* (London: Jonathan Cape, 1944).
Potter, David, *The Talyllyn Railway* (Newton Abbot: David St John Thomas, 1990).
Russell, R.S., *Robert Aickman: An Attempted Biography* (North Yorkshire: Tarturus Press, 2023).
Thirsk, Joan (ed.), *The English Rural Landscape* (Oxford: Oxford University Press, 2000).
Thompson, E.P., *The Making of the English Working Class* (London: Victor Gollancz, 1963).
Thompson, Flora, *Lark Rise to Candleford* (Oxford: Oxford University Press, 1945).
Thurston, E. Temple, *The Flower of Gloster* (London: Williams and Norgate, 1911).
Warner, Pat, *Lock Keeper's Daughter* (Shepperton: Shepperton Swan Ltd, 1986).
White, Alan, *The Worcester and Birmingham Canal* (Studley: Brewin Books, 2005).
Whitehouse, Michael, *Talyllyn Pioneers* (Stourport: Wilderness Enterprises, 2015).
Willan, T.S., *The Navigation of the Weaver in the Eighteenth Century* (Manchester: Chetham Society, 1951).
Willan, T.S., *River Navigation in England 1600–1750* (London: Frank Cass & Co. Ltd, second edition, 1964).
Willan, T.S., *The Inland Trade* (Manchester: Manchester University Press, 1976).
Williams, Raymond, *The Country and the City* (London: Chatto & Windus, 1973).
Williams, Raymond, *Politics and Letters* (London: Verso, 1979).
Williams, Raymond, *Border Country* (Cardigan: Parthian Books, 2005).
Wright, Patrick, *On Living in an Old Country* (London: Verso, 1985).
Wright, Patrick, *Tyneham: the Village that Died for England* (London: Faber & Faber, revised edition, 2002).
Young, Francis Brett, *The Young Physician* (London: Collins, 1919).

# NOTES

*Chapter 1 and Interlude*
1 Ian Mackersey, *Tom Rolt and the Cressy Years* (London: M. and M. Baldwin, 1983); David Bolton, *Race Against Time: How Britain's Waterways Were Saved* (London: Methuen, 1990).
2 Joseph Boughey, 'From Transport's Golden Ages to an Age of Tourism: L.T.C. Rolt, Waterways Revival and Railway Preservation in Britain, 1944–54', *Journal of Transport History*, Vol. 34, Issue 1 (June 2013).
3 L.T.C. Rolt, *Narrow Boat* (London: Eyre & Spottiswoode, second edition, 1947), p. 30.
4 Rolt, *Narrow Boat*, preface to second edition and p. 142.
5 David Matless, *Landscape and Englishness* (London: Reaktion Books, second edition, 2016).
6 Rolt, *Landscape with Machines* (London: Longman, 1971), p. 171.
7 www.ltcrolt.org.uk (accessed 28 February 2024).
8 Victoria Owens, *The Life of L.T.C. Rolt: Where Engineering Met Literature* (Barnsley: Pen & Sword History, 2024).
9 Raymond Williams, *Politics and Letters* (London: Verso, 1979), p. 23.
10 The idea of 'interludes' has been adopted from David Matless, *In the Nature of Landscape: Cultural Geography on the Norfolk Broads* (Chichester: Wiley Blackwell, 2014).
11 Patrick Wright, *Tyneham: The Village that Died for England* (London: Faber & Faber, revised edition, 2002), Chapters 10 and 12.
12 Patrick Wright, *On Living in an Old Country* (London: Verso, 1985).
13 Matless, *Landscape and Englishness*, p. 78.
14 *Ibid.*, p. 80.
15 Raymond Williams, *Politics and Letters* (London: Verso, 1979), pp. 21–23.
16 *Ibid.*, p. 21.
17 *Ibid.*, p. 67.
18 Raymond Williams, *Border Country* (London: Chatto & Windus, 1960), p. 75. The words are spoken by a character in this 1960 novel, but seem to express Williams' view.
19 Rolt, *Narrow Boat*, p. 18.
20 *Ibid*, p. 18.
21 Kate Tiller, 'Hook Norton – An Open Village' in Joan Thirsk (ed.), *The English Rural Landscape* (Oxford: Oxford University Press, 2000).

## Chapter 2 and Interlude

1 Charles Hadfield, foreword to *Narrow Boat*. On a personal note, in researching *Charles Hadfield: Canal Man and More* (Stroud: Sutton Publishing Ltd, 1998), I asked Charles whether his comment was not perhaps a somewhat exaggerated tribute to an old friend, but his reply was trenchant, 'I meant every word' (unpublished correspondence).
2 Ian McNeil, 'The Publication of Works on the History of Engineering and Technology', *Transactions of the Newcomen Society* (1979–80), 51, pp. 219–25.
3 L.T.C. Rolt and Patrick Whitehouse, *Lines of Character* (London: Constable, 1952).
4 John Bate, *The Chronicles of Pendre Sidings* (Chester: RailRomances, 2002), p. 137.
5 Rolt, *Narrow Boat*, p. 167.
6 This was Beech's Dock: a plaque commemorates the conversion of *Cressy* there in 1929.
7 Rolt, *Landscapes with Machines*, p.113.
8 *Ibid.*, p. 215.
9 There was an earlier unpublished novel, *Strange Vista*, and among juvenilia was *The Railway*, discussed in Victoria Owens, '*The Railway*: an Unknown Early Work by L.T.C. Rolt', *Journal of the Railway and Canal Historical Society*, Vol. 41, Part 1, No. 246 (March 2023).
10 Rolt, *Narrow Boat*, p. 12.
11 Rolt, *Landscapes with Machines*, p. 225.
12 E. Temple Thurston, *The Flower of Gloster* (London: Williams and Norgate, 1911).
13 This was Charles Woodhouse, listed as a 'coal merchant' on the 1939 Register. The County Arms, which was developed on the site of the former Union Inn, was closed in 2002 and has since been converted into apartments.
14 Rolt, *Narrow Boat*, p. 70.
15 *Ibid.*, p.82.
16 *Ibid.*
17 *Ibid.*, pp .128–29. The tug would be withdrawn in 1954.
18 According to painter Kay Andrews' website, these were painted by Matilda Woodhouse at Buckby. See kayscanalcraftyarts.co.uk/buckby-watercans (accessed 24 February 2024).
19 I owe these points to an extremely helpful private email, dated 1 January 2004, from Tom Foxon, who was involved with carrying narrow boats from 1950. He stressed that these were only minor criticisms of what remains an admirable book.
20 This would not apply after the founding of the Inland Shipping Group in 1971–72.
21 Harry Hanson, *The Canal Boatmen 1760–1914* (Manchester: Manchester University Press, 1975).
22 Notably by Christopher M. Jones, who presented many findings at a lecture to the Boat Museum Society, Ellesmere Port in 2004.
23 Rolt, *Narrow Boat*, p. 36.
24 *Ibid.*, p. 195.

25 Unpublished log of *Cressy*, 3, 26 and 27 August 1939 and 29 November 1939. These encounters were almost certainly reflected in discussions in *The Inland Waterways of England* (q.v.).
26 One exception was Herbert Mackey of Warrington, who converted the former passenger vessel *Duchess Countess* and used this to travel to the canal to Llangollen. Rolt encountered him in 1949 when the boat had been pulled out and settled on the canalside near Ellesmere.
27 See Joseph Boughey, 'Early Pleasure Boating on the Shropshire Union Canal', *Waterways Journal*, Vol. 7 (2005).
28 For Norman Anglin, see Joseph Boughey, 'Norman Anglin, Predecessor to L.T.C. Rolt', *Waterways Journal*, Vol. 15 (2013), and 'Norman Anglin: A Postscript 2020', *Waterways Journal*, Vol. 23 (2021).
29 Rolt, *Narrow Boat*, p. 93.
30 These included Rolt himself about Thames pleasure boating, in *The Thames from Mouth to Source* (London: Batsford, 1951).
31 Rolt, *Landscape with Canals* (Stroud: Sutton Publishing, 1977), pp. 72–73.
32 Rolt, *Narrow Boat*, p. 60.
33 *Ibid.*, p. 194.
34 *Ibid.*, p. 80.
35 The National Archives, MT52/109, Frank Pick, Report on Canals and Inland Waterways to the Minister of War Transport, unpublished, 1941.
36 Pick Report, p. 36.
37 *Ibid.*, p. 21.
38 *Ibid.*, p. 59.
39 Rolt, *Narrow Boat*, p. 141.
40 *Ibid.*, p. 143.
41 Others in the IWA, notably Charles Hadfield, would advocate enlargement, while the proposals for a Grand Contour Canal, a ship canal at the 310ft contour, would be favoured for a long time. Joseph Boughey and Charles Hadfield, *Charles Hadfield: Canal Man and More*, pp .80 and 97.
42 Rolt, *Narrow Boat*, p. 165.
43 The registers are now in the Cheshire Record Office.
44 *Staffordshire Sentinel*, 28 March 1956.

## Chapter 3 and Interlude
1 Rolt, *Landscape with Canals*, p.50.
2 *Ibid.*, p. 58.
3 The term 'fringe' is used not only because these figures were outside the mainstream of the Right, but also because they remained uninfluential outside various writings.
4 John Michell, Foreword to Edward Abelson (ed.), *A Mirror of England* (Hartland: Green Books, 1988), pp. x and xi.
5 Arthur Bryant, *The Spirit of Conservatism* (London: Methuen, 1929), pp. 74–75.
6 Matless, in *Landscape and Englishness*, has commented on the later Prime Minister Stanley Baldwin's use of similar language, but stressed that Baldwin's view was

static. Baldwin celebrated the rural but proposed no action. Bryant had read drafts of Massingham's book, *The Curious Traveller* (London: Collins, 1950).
7 Lymington and Gardiner had been members of the English Mistery, which advocated a new feudal order.
8 On Wallop, see Philip Conford, 'Organic Society: Agriculture and Radical Politics in the Career of Gerard Wallop. Ninth Earl Of Portsmouth' in *Agricultural History Review*, Vol. 53, Part 1 (2005), pp.78–96.
9 Richard Moore-Colyer, 'A Voice Clamouring in the Wilderness: H.J. Massingham (1888–1952) and Rural England' in *Rural History*, Vol. 13, Part 2 (2002).
10 H.J. Massingham, *The English Countryman* (London: Batsford, 1942). The wording in 'Kinship of Husbandry' diverges from the usual 'Kinship in Husbandry'.
11 He claimed to have later joined Plaid Cymru, but during the period when it was led by the conservative figure Saunders Lewis.
12 Boughey and Hadfield, *Charles Hadfield: Canal Man and More*, pp.14–15. Klingender was author of *Art and the Industrial Revolution* (London: Noel Carrington, 1947). See letter from Aickman to Rolt, 5 December 1947 (National Archives under PRO 30/82/1-7).
13 Sonia Rolt kindly allowed me to see these on a visit to Stanley Pontlarge in August 2004. Letters cited from Massingham below are from this collection.
14 Rolt, *Narrow Boat*, p. 72.
15 Rolt, *Landscape with Canals*, p. 75.
16 Rolt, *High Horse Riderless* (London: George Allen & Unwin, 1947), pp.14–15.
17 *Ibid.*, p. 17.
18 *Ibid.*, p. 18.
19 Rolt, *Winterstoke* (London, Constable, 1954), p. 46.
20 *Ibid.*, p. 241.
21 Rolt, *High Horse Riderless*, p. 121.
22 *Ibid.*, p. 122.
23 *Ibid.*, p. 126.
24 *Ibid.*, p. 118.
25 *Ibid.*, p. 171.
26 This was a point stressed, given the increased population in Britain which would draw upon food resources, by Aldous Huxley in *Ends and Means* (London: Chatto & Windus, 1937), p. 15.
27 Matless, *Landscape and Englishness*, Chapter 3.
28 For an early pejorative use of the term, related to the then new BBC, see *Punch*, 23 December 1925.
29 An early defence of the use of 'middlebrow' is Nicola Humble, *The Feminine Middlebrow Novel, 1920s to 1950s: Class, Domesticity and Bohemianism* (Oxford: Oxford University Press, 2001).
30 Rolt, *High Horse Riderless*, p. 86.
31 Rolt, *Landscape with Machines*, p. 168.
32 *Ibid.*, p. 203.
33 Harry Blamires, *Twentieth-Century English Literature* (London: MacMillan, 1982), p. 120.

34 W. Reid Chappell, *The Shropshire of Mary Webb* (London: Cecil Palmer, 1931).
35 Francis Brett Young, *The Young Physician* (London: Collins, 1919), p. 485.
36 *Ibid.*, p. 233.
37 *Ibid.*, p. 235.
38 *Ibid.*, p. 236.
39 Raymond Williams, *The Country and the City* (London: Chatto & Windus, 1973), p. 253.
40 Kristin Bluemel, 'Beyond Englishness: the Regional and Rural Novel in the 1930s', in Benjamin Kohlmann and Matthew Taunton (eds), *A History of 1930s British Literature* (Cambridge: Cambridge University Press, 2019), p. 18.
41 S.P.B. Mais, *This Unknown Island* (London: Putnam, 1932), p. vii.
42 Ronald Blythe, *A Writer's Day Book* (Nottingham: Trent Editions, 2006), pp. 12–13.
43 *Ibid.*, p. 15.
44 Richard Llewellyn, *How Green was my Valley* (London: Michael Joseph, 1939). The film of the same name was released in 1941.
45 Letter from H.J. Massingham to Rolt, 17 December 1943.
46 Flora Thompson, *Lark Rise to Candleford* (Oxford: Oxford University Press, 1945).
47 H.J. Massingham, Introduction to Thompson, *Lark Rise to Candleford*, p. 10.
48 Raymond Williams, *The Country and the City* (London: Chatto & Windus, 1973), p. 1.
49 John Michell, Foreword to Edward Abelson (ed.), *A Mirror of England: Anthology of Writings by H.J. Massingham* (Hartland, Devon: Green Books, 1988), pp. 7–8.
50 Rolt, *Landscape with Machines*, p. 3.
51 *Ibid.*, p. 7.
52 *Ibid.*, pp. 31–32.
53 *Kelly's Directory of Cheshire*, 1914 edition, indicates this for Watergate Street.
54 Nikolaus Pevsner and Edward Hubbard, *The Buildings of England: Cheshire* (London: Penguin, 1971), pp. 130–31.
55 *Cheshire Observer*, 6 April 1957.
56 Rolt, *Landscape with Machines*, p. 30.

## Chapter 4

1 The nearest to a full account of the first twenty years is David Bolton's *Race Against Time* (London: Methuen, 1990). This is less inaccurate than Robert Aickman's own *The River Runs Uphill* (North Yorkshire: Tarturus Press, 2014) but is similarly partisan.
2 These include Rolt's own account in *Landscape with Canals*, Chapters 5–8. Much is covered (or evoked) in the collected correspondence between Rolt and Aickman, now in the National Archives under PRO 30/82/1-7.
3 Aickman's precise involvement in the Richard Marsh Literary Agency is unclear. It operated largely between 1941 and 1946. See R.S. Russell, *Robert Aickman: An Attempted Biography* (North Yorkshire: Tarturus Press, 2023), Chapter 6.

4 Massingham seems to have joined the IWA, but to have played no active part. He died in 1952 but was still a (passive) member late in 1948.
5 Robert Aickman, *The River Runs Uphill*, pp. 24–25.
6 This idea was put forward from the early 1950s onwards – in effect, less-used waterways would be placed under a public/private trust, with extensive enthusiast and voluntary involvement. Some may have seen this fulfilled in the Canal and River Trust, established with charitable status to manage the former nationalised waterways from 2012.
7 IWA *Bulletin* 9 (n.d. [1947?]).
8 Letter from Charles Hadfield to Tom Rolt, 28 January 1946, quoted in *Charles Hadfield: Canal Man and More*, p. 14.
9 It is likely that Aickman was exaggerating when he said, 'It has long occurred'; he was influenced by a visit to the Stratford Canal at a time when this was decrepit.
10 It would later focus upon anti-flouridisation and opposition to the EEC. See James Hinton, 'Militant Housewives: the British Housewives League and the Attlee Government', *History Workshop Journal*, Vol. 38, Issue 1, Autumn 1994, pp.129–56.
11 There seems to have been very little membership in Wales and only one active member in Scotland. In 1949, when faced with the proposed closure of the Rochdale Canal, it was pointed out that there were no IWA members 'on the spot'.
12 *Gloucester Journal*, 23 July 1927.
13 *Western Mail*, 20 July 1927.
14 This would be privately financed until the formation of the Glyndebourne Festival Society in 1952 – it was possibly closer to the example of the Talyllyn Railway Preservation Society or Lower Avon Navigation Trust than the general canal system. It began to put on regular festivals, which may have given Aickman the idea of starting a canal-based festival.
15 Letter from Aickman to Rolt, 1 March 1946. He seems to have been a largely passive member.
16 *Hampshire Telegraph*, 17 September 1937.
17 *Hendon & Finchley Times*, 5 January 1940.
18 Notably the Norfolk Wherry Trust, founded in 1949, and the Humber Keel Trust.
19 *Charles Hadfield: Canal Man and More* (Stroud: Sutton Publishing, 1998), Chapters 3–5.
20 *Ibid.*, p.35.
21 Letter from Rolt to Aickman, 28 January 1946.
22 Letter from Charles Hadfield to Rolt, 20 January 1946.
23 Letter from Rolt to Aickman, 23 January 1946.
24 Until 1948, there was an existing Canal Association, so this title could not be used. Between 1912 and 1921 there had been a Waterways Association, which campaigned for the development of enlarged waterways.
25 Letter from Rolt to Aickman, 12 February 1946. The Pure Rivers Society had been founded in 1926 but was floundering in wartime.

26 Letter from Aickman to Rolt, 9 February 1946.
27 *Ibid.*, 1 August 1946, and letter of 27 June 1946.
28 Letter from Rolt to Aickman, 27 February 1947.
29 *Ibid.*, March 1948.
30 *Ibid.*, 18 February 1949.
31 *Ibid.*, 8 December 1949. Aickman, in *The River Runs Uphill*, inaccurately dated Rolt's announcement to a meeting in January 1950.
32 Letter from Rolt to Aickman, 8 December 1949.
33 *Ibid.*
34 The Waterways Archive, CRT/BW/95/279.
35 Letter from Rolt to Aickman, 6 February 1950.
36 IWA *Bulletin* 20 (June 1949).
37 IWA *Bulletin* 12 (n.d. [1948]).
38 The docks, most formerly railway owned, were the more significant assets. Management would largely be split after 1955.
39 Aickman found that one leading NAIWC member felt that narrow boats were finished and only trade on larger waterways had a future. Letter from Aickman to Rolt, 3 February 1949.
40 S.E. Barlow joined the IWA Council in the summer of 1947 and resigned in 1950. IWA *Bulletin* 7 (n.d. [1947]) and 25 (June 1950).
41 IWA *Bulletin* 13 (n.d. [1948]). In a letter to Aickman on 10 June 1948, Rolt described the Severn Company manager as being keen to run down and close the Worcester & Birmingham.
42 Rolt, 'The Future of the Canals', *The Field*, 10 June 1944.
43 Rolt, *Landscape with Canals*, p. 83.
44 IWA *Bulletin* 8 (n.d. [late 1947]).
45 The 'Navigation Hints' were announced in IWA *Bulletin* 6, June 1947.
46 IWA *Bulletins* 8 and 11. Wyatt began his hiring operation from Stone in Staffordshire with four boats. His descendants still run this company from Stone.
47 IWA *Bulletin* 23 (December 1949), publicised Abbott's plans. For Abbott, see David Henthorn Brown and Angela Clark, 'Holt Abbott – a pioneer in canal cruiser design and hire boat operation and his surviving craft', *Waterways Journal* 19, 2017.
48 Preparation of this leaflet was announced in IWA *Bulletin* 6 (June 1947).
49 Angela Rolt was heavily involved in work for the exhibition. The exhibition included Tom's cousin, Stephen Taylor MP, who lent his model narrow boat. IWA *Bulletin* 7 (n.d. [1947]).
50 IWA *Bulletin* 1 (November 1946).
51 IWA *Bulletin* 14 (n.d. [1948]).
52 Joseph Boughey, 'The Decline of the Derby Canal: A Case Study in Independent Ownership and Decline', *Journal of the Railway and Canal Historical Society*, Vol. 31, Pt 3, No. 155 (July 1993); IWA *Bulletin* 14 (n.d. [1948]).
53 Rolt, *Landscape with Canals*, p. 146.
54 Little would be achieved beyond exhortation until the 1960s, with the formation of the Canal Transport Marketing Board, and later, the founding of the Inland Shipping Group in 1971–72.

55 Letter from Rolt to Aickman, 12 September 1949.
56 IWA *Bulletin* 21 (August 1949) and 22 (October 1949).
57 Jodie Matthews, *The British Industrial Canal* (Cardiff: University of Wales Press, 2023), p. ix.
58 IWA *Bulletin* 25 (June 1950), reported that Barwell intended to commence voluntary working parties on the Lower Avon, while the *Gloucestershire Echo*, 21 July 1950, reported that several working parties had taken place on the Avon, in particular on Strensham Lock. Owen Prosser of Sutton Coldfield reported on one working party in Bulletin 27 (n.d. [early 1951]).
59 Letter from Rolt to Aickman, 14 July 1950. Tom Rolt had told Aickman earlier, 'I think the major difference of view between us has always been over the idea of the Association as a "general cultural organisation"'. Letter from Rolt to Aickman, 24 December 1949.
60 Letter from Rolt to Aickman, 19 July 1950.
61 Letter from Aickman to A.S. Cavender, 16 October 1950.
62 Rolt, *Landscape with Canals*, p. 173.
63 Letter from Rolt to Aickman, 29 December 1947.
64 *Ibid.*, 18 February 1949, and from Aickman to Rolt, 22 February 1949. Traffic at the eastern end of the Rochdale had ended around 1940; this was inhibited by a break of gauge between it and the Calder & Hebble at Sowerby Bridge.
65 Rolt, *Landscape with Canals*, p. 158.

## Chapter 5 and Interlude

1 Ruth Delany, *Ireland's Inland Waterways* (Belfast: Appletree Press, 2004), p. 98.
2 Rolt, *Landscape with Canals*, p.77; letter from Aickman to Rolt, 26 September 1946.
3 Letter from Rolt to Aickman, 29 October 1945.
4 Rolt, *Landscape with Machines*, p. 91.
5 One of his earliest youthful writings was a fictional story about railways, discussed in Victoria Owens, 'The Railway: an unknown early work by L.T.C. Rolt', *Journal of the Railway and Canal Historical Society*, Vol. 41, Part 1, No. 246, March 2023.
6 He later wrote an account of lesser railways throughout the British Isles in *Lines of Character*, (London: Constable, 1952), Ch. 4.
7 https://www.rte.ie/archives/exhibitions/emaon-de-valera/719124-address-by-mr-de-valera/ (accessed 1 March 2024).
8 Patrick Wright, *On Living in an Old Country* (London: Verso, 1985), pp. 83–84, quoting Peter Scott's broadcast.
9 Tim Pat Coogan, *Eamon de Valera: The Man Who was Ireland* (New York: Barnes & Noble, 1993), p. 671.
10 Seamus Deane, *A Short History of Irish Literature* (London: Hutchinson Education, 1986), p. 211.
11 R.F. Foster, *Modern Ireland 1600–1972*, (London: Allen Lane, 1988), pp. 537–38.
12 This service, from Fishguard to Waterford, would be withdrawn in the 1950s.
13 Rolt, *Green & Silver*, p. 16.
14 *Ibid.*, p. 249.

15 *Ibid.*, p. 64.
16 *Ibid.* As traffic was to be withdrawn in 1960, this would not come about.
17 *Ibid.*, pp. 136–37.
18 *Ibid.*, pp. 23–24.
19 *Ibid.*, p. 79.
20 *Ibid.*, p. 47.
21 *Ibid.*, p. 49.
22 *Ibid.*, p. 117.
23 *Ibid.*, p. 118–19.
24 *Ibid.*, p. 159.
25 *Ibid.*, pp. 212–13.
26 *Ibid.*, p. 154.
27 *Ibid.*, pp. 161–62.
28 *Ibid.*, p. 162.
29 *Midland Counties Advertiser*, 14 March 1946.
30 The SDA seems to have been a very small body; E.P. Flynn, the local Grand Canal agent, was its secretary.
31 Rolt, *Green & Silver*, p. 99.
32 *Ibid.*, Ch. 7.
33 Declan Kiberd, *Inventing Ireland* (London: Jonathan Cape, 1995), pp. 492–93.
34 Rolt, *Green & Silver*, p. 104.
35 *Ibid.*, p. 106.
36 *Ibid.*, p. 239.
37 On the late Dick Stanley, see Robert and Finola Harris' excellent website https://roaringwaterjournal.com/2016/10/30/the-fiddle-makers-ghost (accessed 1 March 2024).
38 For instance, '"The Ireland That I Would Have": De Valera & the Creation of an Irish National Image' by Michele Dowling, *History Ireland* (Spring 1988), pp. 37–42. For Muintir na Tire, see Eion Deveraux, 'The Lonely Furrow: Muintir na Tire and Irish Community Development 1931–1991', *Community Development Journal*, Vol. 28, No. 1 (January 1993).
39 *Catholic Standard*, 30 December 1949.
40 Rolt, *Landscape with Figures*, p. 91.
41 Rolt, *Green & Silver*, p. iv.
42 Rolt, *Landscape with Canals*, p. 32.
43 *Ibid.*, p. 34.
44 Jack Warner's youngest daughter, Pat wrote an account of her childhood at Tardebigge: Pat Warner, *Lock Keeper's Daughter* (Shepperton: Shepperton Swan Ltd, 1986).
45 Pat Warner, *Lock Keeper's Daughter*, pp. 95–96.
46 Alan White, *The Worcester and Birmingham Canal* (Studley: Brewin Books, 2005), p. 324.

## Chapter 6 and Interlude

1. It has to be recorded that all craftspeople featured by Rolt were male. Whether there were craftswomen whom Tom Rolt could have featured is unclear.
2. Rolt, *Worcestershire* (1949), p. xiii.
3. *Ibid.*, p. ix.
4. *Tewkesbury Register*, 8 February 1941; *Evesham Standard*, 19 December 1914.
5. Rolt, *Worcestershire*, p. ix.
6. *Ibid.*, p. xvi.
7. *Ibid.*, p. xiii.
8. *Ibid.*, p. 283.
9. *Ibid.*, p. 283.
10. *Ibid.*, p. 21.
11. *Ibid.*, p. 286.
12. Obituary in *Birmingham Daily Post*, 9 January 1962.
13. *Gloucestershire Echo*, 12 March 1946.
14. Rolt, *Worcestershire*, p. 230.
15. *Ibid.*, p. 231.
16. H.J. Massingham, *Where Man Belongs* (London: Collins, 1946), pp. 217–18.
17. Rolt, *Worcestershire*, p. 264.
18. *Ibid.*, p. 267.
19. David Higham, *The Priests and People of Harvington* (Leominster: Gracewing, 2006), p.78. The Friends would be disbanded after the 1955 AGM, when Hodgkinson was 83.
20. *Green & Silver* records that Richard Hodgkinson and his wife Marten boarded at Tullamore and went through the Royal Canal, leaving the Rolts on the River Boyle.
21. Rolt, *Worcestershire*, pp. 156–57.
22. *Ibid.*, p. 249.
23. *Ibid.*, p. 111.
24. *Ibid.*, p. 274.
25. By 1952 there were only three nail-makers left in Bromsgrove. *Birmingham Daily Post*, 13 August 1952.
26. Rolt, *Worcestershire*, p. 122.
27. Rolt, *The Inland Waterways of England* (London: George Allen & Unwin, 1950), pp. 115–16.
28. *Ibid.*, p. 116.
29. *Ibid.*, p. 119.
30. An example of this was the Royal Canal in Ireland, upon which traffic was animal-drawn until it ended in 1951.
31. Rolt, *The Inland Waterways of England*, p. 138.
32. Harry Hanson, *The Canal Boatmen 1760–1914* (Manchester: Manchester University Press, 1975).
33. H.J. Massingham, *Remembrance* (London: Batsford, 1941), pp. 54–59.
34. Rolt, *The Inland Waterways of England*, p. 174.

35 *Ibid.*, p. 175.
36 *Ibid.*, p. 177.
37 *Ibid.*, p. 175. For Romani decoration, much of it based on carving rather than paintwork, see David Smith, 'Gypsy aesthetics, identity and creativity: the painted wagon', in Thomas Acton and Gary Mundy (eds), *Romani Culture and Gypsy Identity* (Hatfield, University of Hertfordshire Press, 1997), Ch. 1.
38 Rolt, *The Inland Waterways of England*, pp. 177–78.
39 *Ibid.*, p. 178.
40 *Ibid.*, pp. 180-81.
41 Belloc's *The Servile State*, published in 1912, had echoes of Cobbett and later would be cited by Friedrich von Hayek in 1945.
42 *Wilts & Gloucestershire Standard*, 18 September 2020.
43 Rolt, *The Inland Waterways of England*, p. 193.
44 *Ibid.*, p. 184.
45 *Ibid.*, pp. 184–85.
46 *Ibid.*, p. 182.
47 While there is no special significance in the date, Rolt stated that this was ten years after the death of Joseph Lyne ('Father Ignatius') at Llanthony Abbey, which the party visited on the way. Lyne died in October 1908, suggesting 1918 or 1919. It could not have been June 1920, since Rolt had started boarding school in May of that year.
48 Rolt, *Landscape with Machines*, p. 20.
49 *Ibid.*, p. 21.
50 *Ibid.*, p. 21.
51 *Ibid.*, p. 22.
52 *Ibid.*, p. 35.
53 His son Tim has kindly informed me that these tended to be less family holidays than ones shared with his mother, Sonia.
54 Rolt, *Landscape with Machines*, p. 49.
55 Rolt, *Landscape with Figures*, pp. 222–23; *Birmingham Daily Post*, 26 January 1955.
56 William Plomer, who edited the diaries, visited the area in 1938 and felt it was little changed. See Peter Alexander, *William Plomer: A Biography* (Oxford: Oxford University Press, 1991), p. 217.
57 H.J. Massingham, *The Southern Marches* (London: Robert Hale, 1952), p. 1.
58 *News Chronicle*, 17 August 1951.
59 H.J. Massingham, *The Southern Marches*, p. 140.
60 Rolt, *Landscape with Machines*, p. 35.

## Chapter 7 and Interlude

1 Examples would include the Newcomen Society and, much later, the Association for Industrial Archaeology, of which Tom Rolt was the founding president.
2 Rolt, *Landscape with Figures*, pp. 129 and 135.
3 In private conversation between Joseph Boughey and Charles Hadfield, 1990s.
4 *Charles Hadfield: Canal Man and More* (Stroud: Sutton, 1998).
5 There is a bibliography, complete to 1977, in W.H. Chaloner and B.M. Ratcliffe (eds), *Trade and Transport: Essays in Economic History in Honour of T.S. Willan*

(Manchester: Manchester University Press, 1977), pp. ix–x. Before his 1965 study of the Don Navigation, Willan's previous waterways publication was in 1951, a study of the Weaver Navigation.

6  C.B. Phillips, 'Obituary: Professor T.S. Willan' in *The Independent*, 22 June 1994. Willan stayed at Manchester after retirement with Emeritus status and remained involved in the 1980s.

7  See, for instance, T.S. Ashton's fierce attack in his review of J.M. Keynes' *General Theory* in the *Manchester Guardian*, 24 February 1936. Many economic historians advised or worked for government in wartime.

8  In 1941, his cousin was a solicitor in Hawes and his brother Clerk to the Aysgarth Rural District Council. T.S. Willan and E.W. Crossley (eds), 'Three Seventeenth-Century Yorkshire Surveys', *Yorkshire Archaeological Society Record Series* Vol. CIV (1941). In later years, Willan used family documents to reconstruct the life of a relative who was a retailer in Kirkby Stephen in the western Dales. He was still recorded as living in Hawes in 1936. I am grateful to my sister, Dr Christine Barnes, for assistance on genealogical questions.

9  By odd coincidence, in 1902 Unwin married Frances, daughter of the Wesleyan minister Reverend Mark Guy Pearse (1842–1930), who was author of *Rob Rat*, a tract against the practice of living-in on canal boats. Frances Mabelle Unwin (1869–1956) would work on a biography of her father. See *Mark Guy Pearse: Preacher, Author, Artist, 1842–1930* (Truro: Epworth Press, 1930).

10  By coincidence, William McConnel would acquire the quarry and build the Talyllyn Railway as the cotton business declined. It is unclear whether Tom Rolt knew anything of this before his interest in the Talyllyn, which is covered in Chapter 8.

11  *Manchester Guardian*, 3 February 1925.

12  Much of this is based on T.A.B. Corley, 'George Unwin: A Manchester Economic Historian Extraordinary', Working Paper, University of Reading: Henley Business School, 2002, online draft https://www.reading.ac.uk/Econ/Econ/workingpapers/emdp.pdf (accessed 1 March 2024). The quotation is from p. 31. See also Negley Harte, *The Study of Economic History* (London: Routledge, 2006).

13  Although Clark's field was the seventeenth century (upon which he had published a book in 1929), he focused mainly on Anglo–Dutch relations.

14  T.S. Willan, *River Navigation in England 1600–1750* (London: Frank Cass & Co. Ltd, 2nd edition, 1964), pp. 15 and 91.

15  Michael Robbins, the first editor of the *Journal of Transport History*, wrote that Jackman's was 'a really astonishing book to come apparently out of the blue' and 'was a remarkable performance, with no predecessors, and for a long time no successors either'. Michael Robbins, 'The Progress of Transport History', *Journal of Transport History*, Third Series, Vol. 12, No. 1 (March 1991), pp. 75 and 76.

16  *The English Historical Review*, Vol. 32, No. 128 (October 1917), pp. 611–13.

17  J.H. Clapham, *A Concise Economic History of Britain: From the Earliest Times to 1750* (Cambridge: Cambridge University Press, 1949).

18  T.S. Willan, *The Navigation of the Weaver in the Eighteenth Century* (Manchester: Chetham Society, 1951).

19 In later works, Willan turned towards the examination of trade, especially international traders, and the organisation of retailing, as in *Elizabethan Manchester*, his final book.
20 Arthur Young (1741–1820) wrote many books on agriculture in various regions and nations, but these extended to commentary on political and economic factors.
21 T.S. Willan, *The Inland Trade* (Manchester: Manchester University Press, 1976), p. 4.
22 *The Economic History Review*, New Series, Vol. 16, No. 3 (1964), p. 572. Oddly, H. Pat White (1920–94), a one-time editor of the RCHS Journal, was also an academic in the field of transport geography. However, unlike Willan, he was willing and able to address audiences of enthusiasts for transport history.
23 *The Economic History Review*, New Series, Vol. 12, No. 2 (1959), p. 300.
24 For more details about Hadfield's research methods and approaches to publications, see Joseph Boughey, 'Charles Hadfield and Waterways History', *Journal of the Railway and Canal Historical Society*, No. 174, Vol. 33, Pt 3 (November 1999), pp. 126–35.
25 E.T. MacDermot, *History of the Great Western Railway* (London: Great Western Railway, two volumes, 1927 & 1931). MacDermot (1873–1950) was interested in Somerset history and in the Great Western Railway, but his background was in law, not history.
26 *Charles Hadfield: Canal Man and More*, p. 8.
27 I would surmise that as a manager in the Civil Service he would not wish to identify too closely with unions that he might have to negotiate with; his Labour connections did not prevent his appointment by a Conservative Minister of Transport to the British Waterways Board, while his expulsion from the IWA stood him in good stead with those who were seen as its opponents.
28 See, for instance, Boughey & Hadfield, *Charles Hadfield: Canal Man and More*, p. 129.
29 Gerard Turnbull, 'From Thames to Titicata: An Appreciation of Charles Hadfield', *Journal of Transport History*, Third Series, Vol. 8, No. 1 (March 1987).
30 'The Engineers of the English River Navigations, 1620–1760', *Transactions of the Newcomen Society*, Vol. XXIX (1953–54 & 1954–55). Most other publications by Skempton up to then were on soil mechanics, in which he had a leading expertise.
31 Telephone conversation with the author, 1990s.
32 Later ones, including his Don study, were published by Manchester University Press, which also published popular works like Dr David Owen's study of the Manchester Ship Canal.
33 Interview with Sonia Rolt for *Cornerstone Magazine*, Vol. 31 No. 1 (2010). Reproduced online at https://ltcrolt.org.uk/articles/sonia-rolt-griffin.pdf (accessed 1 March 2024).
34 Rolt, 'The History of the History of Engineering', *Transactions of the Newcomen Society*, Vol. 42, No. 1 (1969), p. 152.
35 W.A. McCutcheon, *Technology and Culture*, Vol. 11, No. 3 (July 1970), p. 435. Oddly, McCutcheon went on to cite Tom Rolt's 'lifetime savoring the pleasures of water travel' and his training as an engineer. Neither attribution was quite accurate.

36 Hadfield did write a study of *Atmospheric Railways*, and wrote much about tramroads, especially those in south Wales, while Willan's work would diverge into studies of international traders and the organisation of retailing and other inland trade.
37 John R. Hume, 'The Rolt Memorial Lecture 2002: Technology as Culture', *Industrial Archaeology Review*, XXV; 1 (2003) provides a general analysis of the background to Rolt's interests, especially those in railways.
38 Rolt, 'The History of the History of Engineering', *Transactions of the Newcomen Society*, Vol. 42, No. 1 (1969), p. 156.
39 Joseph Boughey, 'L.T.C. Rolt: Waterways Historian', in *Journal of the Railway and Canal Historical Society*, No. 174, Vol. 33, Pt 3 (November 1999), pp. 140–49.
40 In *Navigable Waterways* (1969), he acknowledged that Jessop was the Rochdale Canal engineer, citing Hadfield's 'study of the company's minute books'.
41 Rolt, *The Inland Waterways of England*, p. 136.
42 The only exception to this might be through writing and other forms of recording.
43 Rolt, *The Inland Waterways of England*, pp. 97–98.
44 Ibid., p. 98.
45 Matless, *Landscape and Englishness*, p. 153.
46 Even more curious is that the latter assertion survived into the posthumous further edition of *Navigable Waterways* by Bryan Marsh (1985). By this time, Helen Harris and Monica Ellis' *The Bude Canal* (1973) had traced and detailed most of the remains of the canal.
47 J.R. Harris, review of *The Steam Engine of Thomas Newcomen* by L.T.C. Rolt; J.S. Allen, *The American Historical Review*, Vol. 83, No. 3 (June 1978), p.720.
48 In an earlier draft, I referred to this as 'fieldwork', but this later term would refer to assiduous recordings with notebook, measurement and photographs, rather than his approach, which relied on intuition and impressions.
49 Gaut was a significant source for *Worcestershire*; with interests rooted in field studies and in natural history, he was not really a historian.
50 Letter from Rolt to Charles Hadfield, 4 July 1945.
51 *Ibid.*, 14 August 1945.
52 *Ibid.,* 14 August 1945.
53 Rolt, *Worcestershire*, p. 101.
54 Richard Dean, 'The Unfinished Leominster Canal', *Journal of the RCHS*, Vol. 32, Pt 2, No. 164 (July 1996), pp. 82–87.
55 Rolt, *Worcestershire*, p. 192. His observation is, of course, based on inaccurate hearsay, which lends weight to railway history enthusiast Charles Clinker's later dictum that the word of the oldest inhabitant would almost always mislead. Only very limited work was carried out on the section west of Leominster.
56 (London: Longman Rees, 1831) – Priestley was by no means always accurate.
57 Willan, *River Navigation*, p. 113.
58 See, for example, E.P. Thompson, *The Making of the English Working Class* (London: Victor Gollancz, 1963).
59 Willan, *River Navigation*, pp. 24–25.
60 T.S. Willan, 'The River Navigation and Trade of the Severn Valley, 1600–1750', *The Economic History Review*, Vol. 8, No. 1 (November 1937), pp. 68–79.

61 *A Survey of Worcestershire* had been reprinted in 1893–98, but it had been written in the early seventeenth century. Thomas Habington was an antiquarian who lived at Hindlip House and much of his work may well have been accurate.
62 J.U. Nef, *The Rise of the British Coal Industry* (London: Routledge, 1932).
63 Willan, 'The River Navigation and Trade of the Severn Valley, 1600–1750'.
64 These, along with a set of maps, were prepared by the canal enthusiast and researcher Philip Weaver (1908–99).
65 Rolt, *Worcestershire*, p. 157.
66 Rolt, *The Inland Waterways of England*, p.72.
67 T.S. Ashton, 'Review of British Canals', *Economic Journal*, Vol. 61, No. 242 (June 1951), p. 410.This was also reflected in a review of *The Canals of Southern England* by H.J. Dyos in *The Journal of Transport History*, Vol. 2.
68 Until 1962, he was a senior civil servant and then a member of the British Waterways Board, during which time he wrote several canal histories. There may have been a perceived conflict of interest.
69 Rolt, *Thomas Telford* (London: Longman Green, 1958), p. xvi.
70 Charles Hadfield, *Thomas Telford's Temptation* (Cleobury Mortimer: M. & M. Baldwin, 1993), p. 189.
71 Rolt, *The Inland Waterways of England*, p. 121.
72 It should be stressed that Rolt might not agree with this; he disdained the contemporary work of social investigators like those of Mass Observation.
73 Rolt, *The Inland Waterways of England*, p. 208.
74 Rolt, *Narrow Boat*, p. 40.
75 *Ibid.*
76 *Banbury Guardian*, 24 May 1973, 7 June 1973.
77 Matthew Armitage, *Forging Ahead* (Yarnton: Windlass Publishing, 2018) provides a detailed history of the yard.
78 *Banbury Guardian*, 10 April 1975.

## Chapter 8
1 Rolt, *Landscape with Machines*, p. 215.
2 *Ibid.*, pp.4–5; Rolt, *Railway Adventure* (London: Constable, 1953) p. 28.
3 Rolt, *Landscape with Machines*, p. 4.
4 Michael Whitehouse, *Talyllyn Pioneers* (Stourport: Wilderness Enterprises, 2015), pp. 42–44.
5 Rolt, *Landscape with Machines*, p. 7.
6 Much of this paragraph follows the account by Bob Gwynne, *Railway Preservation in Britain* (Oxford: Shire Publications, 2011), Chapter 2. The Patent Office Museum, as it was later called, became absorbed into the Science Museum.
7 *Yorkshire Post and Leeds Intelligencer*, 16 May 1927.
8 W.J.K. Davies, *The Ravenglass & Eskdale Railway* (Newton Abbot: David & Charles, 2nd edition, 1981).
9 *Ibid.*, Chapter 4. Katie was moved to the Llewellyn Miniature Railway at Southport in 1918.
10 *Ibid.*, Chapter 6.

11 Alan Holmes, *Talyllyn Revived* (Tywyn: Talyllyn Railway, 2009), p. 22.
12 Owen Prosser founded the Railway Development Association in 1950.
13 *Welsh Gazette*, 19 April 1900.
14 *Aberystwyth Observer*, 13 October 1910.
15 Henry Haydn Jones (1863–1950, knighted in 1937) would be MP for Merioneth until 1945; he was a member of Merionethshire County Council from 1889 onwards.
16 *Cambrian News*, 16 February 1912.
17 Rolt, *Railway Adventure*, p.15. Curiously, Rolt does not mention that the transfer was to Jones' ownership, and that was the era that he and others were trying to conserve. There had been earlier arrangements for passengers to join the train at the 'Slate Wharf' – see, for instance, *Towyn-on-Sea and Merioneth County Times*, 24 August 1904.
18 A.J. Mullay, *Railways for the People: The Nationalisation of Britain's Railways in 1948* (York: Pendragon, 2006), p. 84.
19 Rolt, *Landscape with Figures*, p. 10.
20 *Ibid.*, p. 7.
21 Simon Bradley, *The Railways: Nation, Network and People* (London: Profile Books, 2015), p. 393.
22 Shouster had also been a candidate for the parliamentary seat of Eskdale, in Cumbria, from 1914 and may well have been aware of the conversion of the Ravenglass & Eskdale Railway. Rolt's remarks were reported in David Potter, *The Talyllyn Railway* (Newton Abbot: David St John Thomas, 1991), p. 58.
23 *Ibid.*, p. 58.
24 Rolt, *Railway Adventure*, pp. 54 and 62.
25 Denis Dunstone, *For the Love of Trains* (London: Ian Allan Ltd, 2007).
26 Rolt, *Landscape with Figures*, pp. 8–10; *Railway Adventure*, pp. 41–42.
27 Rolt, *Landscape with Figures*, p. 1.
28 It is claimed to be the first of its kind in the world.
29 Rolt, *Railway Adventure*, p. 147.
30 Potter, *The Talyllyn Railway*, p. 136, suggests that Tom would have been opposed to the use of army units because of his aversion to militarism.
31 Rolt, *Railway Adventure*, p. 53.
32 *Ibid.*, pp. 137–38.
33 *Ibid.*, p. 84.
34 *Ibid.*, p. 85.
35 Rolt, *Landscape with Figures*, p. 19.
36 Rolt, *Railway Adventure*, p. 83.
37 R. Merfyn Jones, *The North Wales Quarrymen, 1874–1922* (Cardiff: University of Wales Press, 1981).
38 For the concept of the *gwerin*, see Christopher Harvie, *A Floating Commonwealth: Politics, Culture and Technology on Britain's Atlantic Coast, 1860–1930* (Oxford: Oxford University Press, 2008), Ch. 5.
39 Rolt, *Railway Adventure*, p. 147.
40 *Ibid.*, p. 130.

41 Ibid., p. 74.
42 J.B. Snell, Introduction to *Railway Adventure* (Thrupp: Alan Sutton Publishing Ltd, 1993 edition), pp. vi–vii.
43 Rolt, *Railway Adventure*, p. 148.
44 Rolt envisaged that the Talyllyn venture might fail in the final paragraph of *Railway Adventure* (p. 150), where he wrote, 'even if the worst should happen and our venture fail for lack of adequate support, at least a gesture has already been made and by no means an idle or fruitless one'.

# INDEX

References to Tom and Angela Rolt, or books by Rolt, have not been included.

Abbey Theatre, Dublin 111
Abbott, Holt, boatbuilder 92
Abergynolwyn 166, 170, 181
Aberystwyth 169
Acton, Cheshire 49, 53–5
Aickman, Ray (Edith) 81, 117
Aickman, Robert 60, 80–99, 117
Alrewas 46
Anglin, Norman 44
Armitage Tunnel 46
Ascendancy 109
Ashton, Thomas S. 143–4, 155
Athlone 101, 104, 107, 112
Avon, River (Warwickshire/Worcestershire) 26, 89, 96, 98, 127, 151–4, 175
Aylestone Hill Tunnel 150–1
Aysgarth 143

Badger, The, pub 52, 56
Baldwin, Stanley 68
Ballinderry 112
Banbury 39, 41,130, 156, 159–65, 171
Banbury Museum 160, 162–3
Banbury Wharf 160
Barbridge 13
Barlaston 37
Barlow, S.E., coal carrier 91, 95
Barnes, Alfred, Minister of Transport 90
Barnsley Canal 88

Barrow River 106
Barwell, Douglas 89, 96
Basingstoke Canal 89–90
Bate, George 117 129–30
Bate, John 36
Bayliss, E.W., manager 91
Beahan, Jack, boatowner 104
Bedworth 44
Belloc, Hilaire 132
Berry, Henry 149
Bewdley 126
Biographical Dictionary of Civil Engineers 148
Birmingham 127, 172, 176
Birmingham, tunnel tug 118, 122, 176
Birmingham Model Railway Club 172
Blaby 39
Black Mountains 21–4, 135–6
Blamires Harry 67
Blisworth 37
Bloomsbury 81
Blythe, Ronald 70
Boderg, Lough 108
Bolton, David 13
Bowes Committee 88
Boyd, James 166–7
Boyle River 110
Brecon Beacons National Park 137–8
Brereton, Amelia 51
Brett Young, Francis 68–9
Bridgewater Canal 130

Brierley Hill 42
Brindley, James 129
British Canals 146–7, 153–4
British Housewives League 84
British Railways 176, 178
British Rolling Mills Ltd 126
British Transport Commission (BTC) 83, 90–1, 94–5, 105
British Transport Historical Records 147, 155
Broads 44
Bromsgrove 116–7, 128
Brunel, Isambard K. 116, 142
Bryant, Arthur 59–60
Buckby 42
Bude Canal 151
Burton 39

Cader Idris 166
Caffrey, Patrick, carrier 106, 114
Caledonian Canal 155
Cambrian Coast railway line 180
Cambridge University 23, 67, 144
Canal Cruising Company 92
*Canals of the North of Ireland, The* 148
*Canals of South Wales and the Border, The* 153
*Canals of Southern England, The* 151, 157
Canterbury Cathedral, Friends of 85
Capel-y-ffin 137
Castle Quay, Banbury 160–4
Castle Wharf, Banbury 160–1
Cavan & Leitrim Light Railway 105
Chaddesley Corbett 127
Chard Canal 151
Cheltenham 168
Cherwell Drive, Banbury 161, 163
Cheshire 39, 49–56,
Chester 16–9, 27, 35, 73–79, 92
Chester General railway station 75
Chesterton, G.K. 132
Church Minshull, Cheshire 39, 46, 50–6, 61
Churche's Mansion, Nantwich 49–50
Circular Line, Grand Canal 110
Cistercians 62, 66

Clapham, John H. 144–5
Clark, George 144
Clonmacnoise 107, 112
Cloondara Mill 101
Cobbett, William 110 131
Coole Park 61, 110
Córas Iompair Éireann (CIÉ) 105
Corley, T.A.B. 144
Corris Railway 173, 176, 183
Council for the Preservation of Rural England (CPRE) 70, 87
Coughton Court 127
Courtney, Lord 143
Coventry Canal 44
Crafted Boats Ltd 117
*Cressy*, narrow boat 35, 37–47, 49–50, 57, 61–2, 76, 87–90, 94, 98, 101, 124, 134
Crewe 39, 49–51
Cromford Canal 94
Cropthorne watergate 154
Cross Lane, Church Minshull 50, 55
Cusop 17, 67

Dane River 145
Daniels, G.W. 144
Darlington 168
David & Charles, publishers 145, 148, 156
de Salis, H.R. 149
de Valera, Eamon 101–4, 112
Dean, Richard 152
Deane, Seamus 103
Derby 168
Derby Canal 89, 94
Derby Motor Boat Club 44, 93
Docks and Inland Waterways Executive (DIWE) 83, 90, 95
Dolgoch 170
Dolgoch, locomotive 182
Droitwich 127
Dublin 106, 108–12
Dudley 127–8
Dudley Tunnel 184
Dunstone, Denis 173

Eaton Hall, Cheshire 100, 168
*Economic History Review, The*, journal 147
Economic History Society 143, 147
Edinburgh University 144
Ellesmere Port 42, 184
Elmley Lovett, Worcestershire 127
Erewash Canal 94
Evesham 36, 96, 154
*Exeter Flying Post*, newspaper 146
Eyes, John 149

Factory Street, Banbury 159–61, 164
Fairbourne Railway 169, 173, 175
Fellows Morton & Clayton, carriers 91
*Field, The*, periodical 124
Ffestiniog Railway 163
Fladbury Lock and Mill 127, 154
Forth & Clyde Canal 100
Foster, R.F., historian 103
Foulkes, William, chair-maker 127
Foxton inclined plane 44
Frankton 37
Friends of Harvington Hall 127

Gaelic League 101
Gailey 13
Gardiner, Rolf 60, 65
Garland, Patrick 173
Gaut, Robert C. 124–5, 149
Gladstone, locomotive 168
Glyn Valley Tramway 100
Glyndebourne Opera 85
*Golden Age of the Canals, The* (film) 14
*Golden Age of Steam, The* (film) 14
Grand Canal 100, 106, 108–9, 114
Grand Canal Company 100, 105–6, 110
Grand Contour Canal scheme, 83
Grand Union Canal, 38, 87, 93, 106
Grand Union Canal Carrying Company 91, 94
Grand Western Canal, 146
Great Western Railway, 23, 89, 147, 171, 176
Greyfriars House, Chester 75, 77

Habington, Thomas 153
Hadfield, Charles 9, 14, 34, 60, 81, 82, 83–4, 85, 86, 97, 142, 146–8, 150–8
Half Moon Inn 138
Halfway House pub 117
Hammond, Barbara and John 156
Harecastle Tunnel 41, 46
Harris, John R 151
Harvington Hall 127
Hatterall hill 136, 139
Hawes, 142–3
Hay-on-Wye 17 135, 137
Heal, Anthony 93
Heal's exhibition 92–3, 95
Hereford 150–1
Herefordshire & Gloucestershire Canal 150–1
Hill, Sir Reginald 90
Hiroshima 64
Hodges, Tommy, boatbuilder 116, 128
Hodgkinson, Henry R. 127
Hodgkinson, Richard and Marten 111, 113, 127
Holmes, Alan 169
Hook Norton 30–2
Hubbard, Edward 74
Huddersfield Canal 89–90
Hudson, W.H., 58
Hungerford 39
Hurcomb, Sir Cyril 90
Huxley, Aldous 67–8
Hyams, Edward 58
Hyde, Douglas 101

Inland Cruising Association 44, 92
Inland Waterways Association (IWA) 50, 60, 80–99,109, 116–7, 121, 147, 181
Inland Waterways Association of Ireland (IWAI) 25, 101, 112–3
Inland Waterways Redevelopment Advisory Committee 184
Institution of Civil Engineers 147
Insull, Tom, blacksmith 117
*Introducing Canals* 151
Ironbridge 19, 184

Jackman, William T., 144–5
Jackson's Lock, Maynooth, 109
Jameson, Fredric 20
Jamestown 106
Jessop, William, engineer 149, 155
Jones, Sir Henry Haydn 161, 167, 170–2, 174
*Journal of Transport History, The* 14, 147
Joyce, Monsignor Timothy 110, 112
*jus naturale* 61

*Katie*, locomotive 100, 169
Keenagh 106
Kennet & Avon Navigation 39, 88–9, 94, 97–8, 171
Kerr Stuarts, manufacturers 36
Key, Lough 110
Kiberd, Declan 111
Killashee 106–7
Killucan 106
Kilrush 104
Kilvert, Francis 70, 137
Kington 152
Kinship in Husbandry 60
Klingender, Francis 60
Knights, L.C. 23
Knill, John, carrier 97

Lancaster Canal 93
Leavis, F.R. 133
Leech, James, carrier 106
Leicester 40–1, 46, 69
Leominster Canal 149, 151–3
Liberal Party 161, 171
Lifford 89
Light Railway Transport League (LRTL) 85
Linton Lock 96–8
Liverpool 92, 149, 169
Llangollen 39, 48, 93–4
Llangollen Canal 90
Llanthony Priory 24, 107, 135–41, 166
Llewellyn, Richard 70–1
*Locomotion*, locomotive 168
Long Crendon 58
Lough Derg Yacht Club 109

Loughborough 39
Lucan, Lord 97
Lugg, River 152
Lymington, Viscount (Lord Portsmouth) 25, 60, 65

McConnel & Kennedy manufacturers, 144
McConnel family 170, 174
McCutcheon, W A 148
MacDermot, E.T. 147
Mackersey, Ian 13
MacNeice, Louis, poet 91
Mais, S.B.P. 39, 70
Mallender, B.A. 94
Mamble 151
Manchester University 143–4
Market Drayton 42
Market Harborough 17, 92–3, 96–7, 129, 172
Marsh, Christopher 94
Mass Observation 21
Massingham, Gertrude S. 58
Massingham, Harold John 25, 37, 57–72, 81, 126, 128, 131, 137–8
Matless, David 20–1, 26, 66, 125–6, 180
Maynooth 108–9, 115
Mellor 144
Middlewich Branch, Shropshire Union Canal 42, 50, 55
Mill Street, Nantwich 15–6
Miller, John 106
Minshull Lock 42
Minshull Mill 52
Minshull Wharf 50, 55
*Modern Tramway*, periodical 172
Morris, William 68, 126
Morris, May 126
Mosley, Oswald 59
Moyasta Junction 107
Muintir na Tíre 112

Nant Gwernol extension 181
Nantwich 14–5, 28, 49–50.
Nantwich Basin 49

Napton 39
Narrow Gauge Railways Ltd 168–9
National Association of Inland Waterways Carriers 91
National Trust 87
National Waterways Conservancy 83, 89
Nef, J.U. 153
*New English Weekly*, periodical 60
Newcomen Society 18, 86, 148, 175
Newnham 151
Newport branch, Shropshire Union Canal 13
Nicholas Street Chester 75
Norbury Junction 13
North Eastern Railway Co. 168
North Kilworth 39
Northgate Street, Chester 75
'Number Ones' 91, 42–3, 105

Oxford Canal 159–64
Oxford University 143–4, 147
Owens, Victoria 19

*Painted Boats* (film) 91
Pandy 22–3
Parker, John F. and Alice 126, 128, 133
Patent Office Museum, 167–8
Pendre Station, Tywyn 174, 178, 182
Penrhyn, Lord 178
Penrhyn Quarries 178
Pensax Tunnel 152
Pershore 44, 125
Pevsner, Nicklaus 74
Phoenix Green garage 37–8, 166
Pick, Frank, and Pick Report 46–8, 91
Plymouth Arms, Tardebigge 118
Pontcysyllte Aqueduct 155
Pooley Hall, colliery 42
Portobello Harbour, Dublin 110
Portumna 110
Potteries 35–9, 68
Pownall, J.F. 83
Prescott Hill Climb 86
Priestley, J.B. 69
Priestley, Joseph 152–3

Prosser, Owen 169, 172, 180
Pugin, A.W., architect 108
Pure Rivers Society 86

Railway and Canal Historical Society 85–6, 147
Railway Club 168
Railway Correspondence and Travel Society 168
*Railway Magazine* 168
Ramblers Association 87
Ravenglass & Eskdale Railway (R&E) 168–9, 174
Rea Aqueduct 151
Reindeer, pub, Banbury 159
Rennie, John, engineer 149
Rhydyronen station 182
Rhyl 169
Richmond Harbour 101
Rimmer, Arthur 172, 176
Ringsend Docks, Dublin 110–1
Rix, Michael 185
Robertstown, Grand Canal 114
Rochdale Canal 89, 98, 149
*Rocket*, locomotive 167
Rockingham House 110
Rolls-Royce factory, Crewe 50
Rolt, Sonia 18, 87, 95, 173
Roma/Romani 29–30, 42, 130–3
Royal Canal 100–1, 106, 108–9, 114–5
Rugeley 46

Salford 44
Sandys, William 153–4
Sankey Navigation 149
Saul 92
Sawley 44
Schumacher, E.F. 65
Scott, Peter 102–3
Severn, River 86, 150, 153
Severn Carrying Company (SCC) 91, 94–5
Shannon, River 100–1, 106, 108–9, 112
Shannon Development Association 110
Shardlow 21
Sharpness 151

Sharpness New Docks company 118
Shiels, George, playwright 111
Shipston-on-Stour 29
Shouster, Sir George 172
Shrewsbury and Hereford Railway Company 153
Shropshire Union Canal 42, 50, 55, 73, 76, 91
Simmons, Jack 185
Skempton, Alec 147–8, 185
Skirrid Inn 137
Smeaton, John, engineer 129
Smiles, Samuel, author 37, 67, 100, 148–9
Smith, George, boatman 87, 95
Smith, Sonia, *see* Rolt, Sonia
Snell, John 179
Somerset & Dorset Railway 179
Southern Railway Company 172
Southnet Tunnel 152–3
Southport 168
Stafford-Harman, Sir Cecil 110
Staffordshire & Worcestershire Canal 13
Stanier, William, engineer 168
Stanley, Dick, violin-maker 112
Stags Head, pub, Swalcliffe 29
Star Inn, Acton 49, 53–4
Stephenson Locomotive Society 172
Stockport 143
Stockton & Darlington Railway 168
Stone 92
Stour, River, Essex and Suffolk 89
Stourport 92, 151
Stratford-upon-Avon Canal 88, 98, 171
Stroudwater Canal 89
Struggler, The, pub 132, 159
Swalcliffe 29, 32–3
Swindon 168 176–7

Tadmarton 29
Talyllyn Lake 166
Talyllyn Railway 93, 98, 100, 105, 134, 161, 166–4, 186
Talyllyn Railway Preservation Society (TRPS) 169, 171–3, 180
Tardebigge 25, 57, 82, 84, 116–24, 128–9, 156
Tardebigge Tunnel 116, 119, 122
Taylor, Dr George 73
Taylor, J.H., boatbuilders 17
Telford, Thomas, engineer 129, 155
Temple Thurston, E., author 39
Territorial Army 176
Thames, River 44, 144
Thames & Severn Canal 132
Thomas, Edward 172
*Thomas Telford's Temptation* 155
Thompson, Flora 71
Tickenhill Manor 126
Tiller, Kate 30–1
Tipton 126
Tixall 42
Tom Rolt Bridge, Banbury 161, 163
Tom Rolt Centenary Rally 16–9
Tooley, George, boatbuilder 128
Tooley, Herbert, boatbuilder 159
Tooley's Yard, Banbury 29, 38, 159–65
Tower Wharf, Chester 17, 76, 78
Trafford Moss 130–1
Traherne, Thomas 68, 138
Tralee & Dingle Railway 113
Tramway & Light Railway Society 85
Tramway Museum Society 85
Transport Act 1947 173
Transport Act 1968 53
Transport & General Workers Union 95
Trent, River 44, 93
Trent & Mersey Canal 46, 92
Trinder, Bill 161, 171–2, 176
Tullamore 107, 110, 113
Tunnel House, pub 132
Tunnel Lane, Lifford 89, 96, 98
Tunnel Tugs 41, 46, 116, 118, 122–3
Tywyn 14, 166, 169–71

Unwin, George 143–4

Vale of Ewyas 22, 24, 135–8
Vale of Rheidol Railway 169
Vaughan, Henry 68, 138
Vesey-Fitzgerald, Brian 124

Victoria County History for
    Worcestershire 124–5
*Village Labourer, The* 156
Vintage Sports-Car Club (VSCC) 37,
    86, 92–3

Walsall 172
Warner, Jack lock-keeper 117
Waterford 104
Watergate Row, Chester 73–5, 79, 136
Watergate Street, Chester 73–5, 79
Waterways to Stratford 153
Weaver, Mr, nail-maker 128
Weaver, River 15, 42, 50–2, 145
Webb, Mary 68
'Welsh Canal' *see* Llangollen Canal
Welsh Highland Railway 100, 169, 172
West Clare Railway 104–5
Weston Point 42
Wharf Station, Tywyn 14, 171, 174,
    181–3
White, H. Patrick 145
Whitehouse, Patrick, 173
Wide Streets Commissioners 111
Widnes 42
Wilkins, John 173, 175

Willan, Thomas S. 25, 129, 142–50,
    153–4, 156–8, 185
Willans, Kyrle 36–8
Williams, Raymond 20–4, 66–7, 69,
    71–2, 171
Williams-Ellis, Clough 70
Winchcombe 29
Winsford 50
Woodhouse, Charles, publican 39
Worcester, tunnel tug 123
Worcester & Birmingham Canal 47,
    94, 129
Worcestershire Folk Museum 126
Working Boaters Sub-Committee, IWA
    95, 98
Wrens Nest Hill, Dudley 127
Wright, Patrick 20–2, 103
Wyatt, Rendel 92, 97
Wye, River 152

Yarranton, Andrew 153–4
Yeats, William Butler 61, 67, 101, 104,
    111
York Railway Museum 168
Young, Arthur 145

## You may also enjoy ...

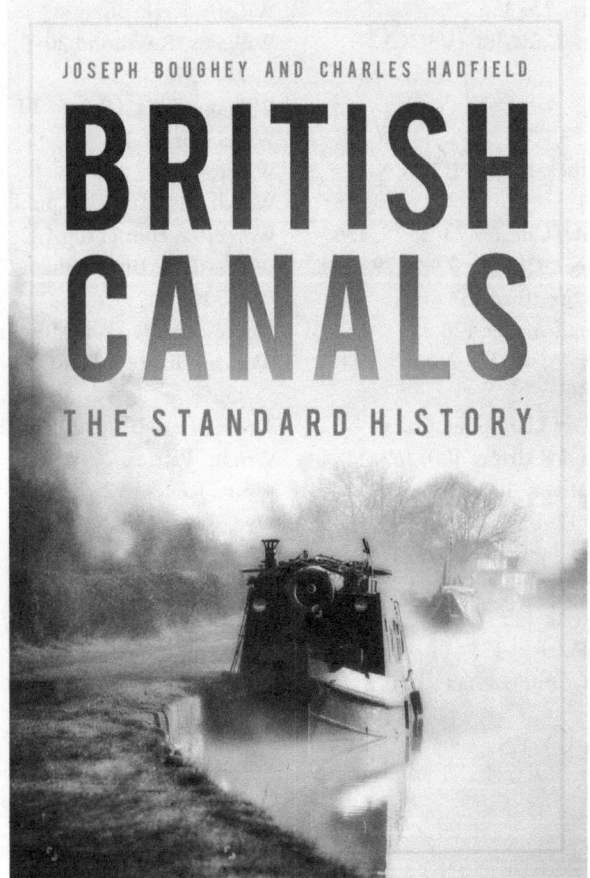

978 0 7509 9937 3

The first edition of *British Canals* was published in 1950 and was much admired as a pioneering work in transport history. For this ninth edition, the many new discoveries and advances in the world of canals around Britain are explored.

The destination for history
www.thehistorypress.co.uk